SCIENCE FAIRS

Ideas and Activities

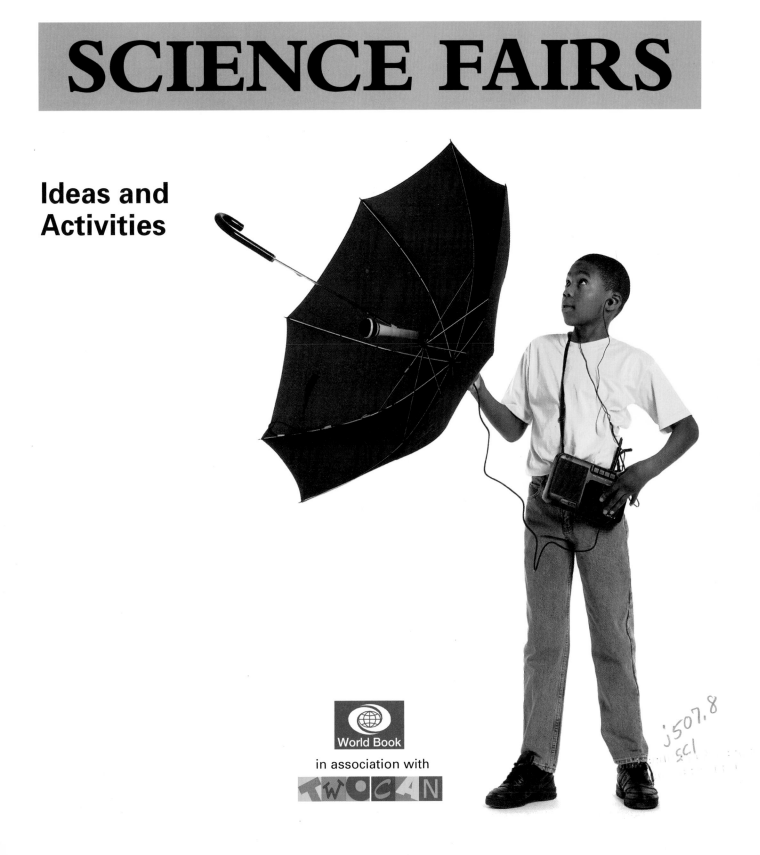

World Book

in association with

TWOCAN

Staff

Publisher Emeritus
William H. Nault

Vice President, Publisher
Michael Ross

Editorial

Managing Editor
Maureen Mostyn Liebenson

Associate Editor
Sharon Nowakowski

Senior Editor
Melissa Tucker

Permissions Editor
Janet T. Peterson

Contributing Editors
Lisa A. Klobuchar
Diana K. Myers
Katie Sharp

Indexers
Pam Hori
David Pofelski

Consultants
Ginger Kranz
Science Specialist
Chicago Children's Museum

Jay Myers
Senior Science Editor
The World Book Encyclopedia

Art

Executive Director
Roberta Dimmer

Art Director
Wilma Stevens

Designer
Sandy Newell

Cover Design
Kristin Nelson

Art Production
Mary-Ann Lupa

Senior Photographs Editor
Sandra Dyrlund

Photographer
Allan Landau

Product Production

Vice President, Production and Technology
Daniel N. Bach

Director, Manufacturing/Pre-Press
Sandra Van den Broucke

Manufacturing Manager
Carma Fazio

Manufacturing Production Assistant
Mike Magusin

Senior Production Manager
Randi Park

Text Processing
Curley Hunter

Proofreaders
Anne Dillon
Carol Seymour

Special thanks to

Kate Asser, Francesca Baines, Peter Clayman, Sarah Davies, John Englefield, Kate Graham, Mike Hirst, Diana Leadbetter, Jacqueline McCann, Helen McDonagh, Paul Miller, Christine Morley, Lisa Nutt, Michael Ogden, Carole Orbell, Leila Peerun, Katherine Senior, Andrew Solway, Robert Sved, Matthew Ward, Claire Watts, Belinda Webster, Jane Wilshir.

**For information on other World Book products,
call 1-800-255-1750, ext. 2238,
or visit our Web site at http://www.worldbook.com**

Published by
World Book, Inc
525 W. Monroe
Chicago, IL 60661
in association with Two-Can Publishing Ltd.

Photos on pages 18-23, 26-35, 37-47, 50-63, 66-73 © Two-Can Publishing Limited, 1997, design © Andrew Haslam. All other photos © World Book, Inc., 1997.

Series concept and original design © Andrew Haslam and Wendy Baker.

Make it Work! is a trademark of Two-Can Publishing Limited.

Printed in the United States of America

2 3 4 5 6 7 8 9 10 02 01 00 99 98   (hc)
1 2 3 4 5 6 7 8 9 10 02 01 00 99 98 97 (sc)

Library of Congress Cataloging-in-Publication Data

Science fairs: ideas and activities.
  p. cm.
Includes bibliographic references and index.
Summary: Ideas for hands-on science fair projects in the areas of space, earth, machines, plants, and time.
  ISBN 0-7166-4498-3 (hc) 0-7166-4497-5 (sc)
  1. Science projects—Juvenile literature. [1. Science projects.
2. Science —Experiments. 3. Experiments.]
Q182.3.W67 1997
507' .8—dc21   97-29786

Introduction

Have you ever thought about using solar energy in your own backyard? How efficient would your solar cooker be? Do you know how to build a model airplane? How far could you make it fly? Suppose you built a model boat, but you wanted to make it faster or strong enough to hold secret cargo–what would you do? Have you ever wanted to get across a river? What type of bridge would you build? Perhaps you wonder about what is in a cave, how a mountain is formed, or how sound travels. Maybe you've even thought about living in space. What would that be like? What will people have to do to survive?

You can use scientific concepts to begin answering these questions right at home–and any one of them could lead to great scientific exploration.

One of the best ways to share your discoveries with others is at a science fair. In this book, you'll find many ideas and activities to get you started on a science fair project. But before you start, be sure to read *The Best of the Fair.* This section will introduce you to science fairs and tell you how to plan for one. Next, begin searching for your topic. Read the titles of the projects listed in the *Contents.* Scan the introductions and questions that precede the project sections–*Space, Dynamic Earth, Machines and Structures, World of Plants, Sound.* And, browse the *Index.* Then, pick an activity and read it. There, you'll find more project ideas and suggestions.

Your project should reflect your original thinking and ideas. Parts of your project will naturally be more difficult than others. Don't be afraid to ask for help or guidance–all science builds on the help and discoveries of others. Use the *Glossary* and *Find Out More* sections at the back of the book. Check out the *Scope and Sequence* to see what skills you'll be reinforcing as you do the various activities.

After all your hard work, you'll have the fun of attending the science fair. There, you will not only learn about your project and others', but you'll also get a chance to talk about your work with teachers and judges. You may even win a prize! But win or lose, the experience will be rewarding. You'll learn a lot about science, communicating with others, and your own abilities and work habits. Plus, you'll know how to better plan for your next school assignment or science fair project!

Contents

48 World of Plants

Explore the vital world of plants. The oxygen that humans breathe comes from plants, and most of the food we eat comes from plants or from animals that eat plants. Study the oxygen cycle, plant life, the diversity of species and their habitats, and the many fascinating uses we have for plants.

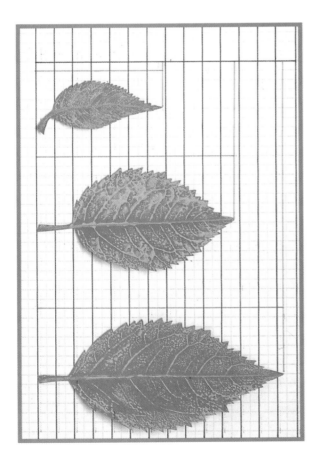

64 Sound

Sound can blow out candles and crack glass. Bats use sound to "see" in the dark and to find their way around. Investigating sound can be fun and exciting. Find out how sound travels through the air, through walls, and around corners. How can it be captured and recorded? What makes a room soundproof?

Names of projects for you to do appear in *italicized* type.

Selecting and researching a topic

Sometimes the hardest step in preparing a science fair project is deciding what to do. Pick a topic that interests you and one that can be explored within the amount of time you have.

Choosing a topic

You can search for inspiration in many places. Read, read, and read some more. Page through your science textbooks. Browse through the science section of your school or public library. Search through encyclopedias. Leaf through science magazines. Read newspapers for current events in science. Scan the Internet.

Look around you. Take a walk, look under the hood of a car, visit a museum or factory, tag along with a volunteer at a science center or museum. Take notes as you explore. If you find yourself thinking, "I wonder what would happen if . . .," you may be close to a good project idea. Write it down!

Think about your hobbies. Do any project ideas come to mind? For example, perhaps you could design a better baseball. Use your imagination.

Strike up a conversation. Talk to professionals at hospitals, universities, government agencies, zoos, museums, and factories, all of which you can find listed in telephone books, industry directories (ask a librarian for help with these), and on the Internet.

Use this book. Do a project that's listed or use one as a springboard to your own idea.

Defining the problem

With your topic in hand, it's time to go back to the library. Look for information that will help you narrow your topic to a specific question or problem. For example, if you are interested in explorers and exploring, you may wonder, "How can hikers keep their food from rotting too fast?"

November 21

World Book Encyclopedia 1998 ed.
Airplane and Lift articles.

fat wings

curved wings

thin wings

"aerodynamics"

drag?

wings

Start your project journal. Take notes from encyclopedias and textbooks, then move to more specialized sources, such as science journals. Consult relevant organizations and agencies. Many have web sites on the Internet. Write to scientists who are conducting research related to your problem. Or, interview experts. Call to ask for a 20-minute interview, write questions ahead of time, and afterward send a thank-you note. This research will also be valuable later, when designing your experiment.

Get started on your bibliography by writing the names of your sources and other important details. Ask your teacher for bibliography guidelines to ensure that you gather the correct information for each type of source (book, CD-ROM, magazine, etc.).

When defining your problem or question, keep in mind the following:

1. A good question is specific but open-ended (does not have a "yes" or "no" answer).

2. Be sure you can complete the project with minimal help. Do you understand the terms, concepts, and methods involved?

3. What equipment or facilities will you need?

4. Do you have enough time to do the work?

5. You may need to chisel away at your original question until you come up with something that's specific enough to manage.

6. Most science fairs have rules about certain kinds of projects, especially ones containing animals and harmful substances. So once you have your project idea, run it by your teacher and parents for suggestions and approval.

7. In addition to satisfying your curiosity, what is the value of finding the answer to your question?

Getting organized

When scientists are faced with a question or problem, they use what they know of the topic to think of a possible answer, or hypothesis. A **hypothesis** is simply an educated guess (not a wild guess), using knowledge gathered from literature and prior experiences. Scientists then create experiments that they can do to prove their hypothesis right or wrong. They conduct their experiment many times, making slight changes each time. From their experiments, scientists collect data, draw conclusions, and create new hypotheses. This process, called the **scientific method,** is the basis of your science fair project.

Before you decide on a hypothesis, come up with a list of factors that could affect the answer to your question. These are called **variables.** For example, if you want to find out how effective different spot removers are in cleaning oiled rags, your variables could include the size of the oil spot, the length of time the spot has been in the cloth, the different spot removers, the amount of spot remover used, and how the spot remover is applied (poured on, rubbed in, etc.). Which variables would you want to test? You could change them—one at a time, of course—to see how each one affects the answer to your question.

As you develop your hypothesis, think about other studies you have read and their findings. Your final hypothesis should be clear and brief, and it should indicate what you plan to do or find out. In other words, your hypothesis should state the purpose of your project. For example, "I think that the spot removers with chemical x will work best because I read...."

You must be able to conduct an experiment that supports or rejects your hypothesis. Make a step-by-step list of what your experiment will entail. Make your plan as simple as possible. The more complicated the design, the more chance for error. Remember, test one variable at a time and note any unknown variables that you will not be able to control but which may affect your outcome. How will you measure the results? List what equipment and materials you will need. Finally, get approval from your teacher or parent.

Scheduling your time

Develop a timetable. Begin by dividing your project into smaller tasks and set a due date for each one. Setting smaller goals will give you a feeling of accomplishment as you meet each deadline. Every few days or so, review your checklist on paper or in your head.

Use this checklist as a guide for developing your own project timetable.

__ Select a topic.
__ Conduct research and propose a question or problem.
__ Develop a hypothesis, purpose, or possible answer to the problem.
__ Plan an experiment to test the answer; get approval and necessary science fair forms and rules.
__ Gather materials and equipment.
__ Conduct the experiment again and again; collect data; take photographs; keep a detailed log.
__ Analyze your data; make calculations; draw diagrams, graphs, and charts. What are the possible meanings of each piece of data?
__ Draw a conclusion. Was your hypothesis correct? What could be explored further?
__ Write your project report.
__ Design and construct your science fair exhibit.
__ Practice your presentation; ask a friend to evaluate it and to ask questions.
__ Attend and enjoy the fair.

Playing the role of scientist

Now put your plan into action. Be sure to work slowly and carefully. You want to avoid making mistakes that may throw off your results.

Conducting experiments

When conducting experiments, you need to include a control. A **control** sets up conditions in an experiment so that you know exactly what variable is responsible for the observed results. If you were conducting the spot remover experiment, for example, your control might be to clean an oily cloth with plain water. The only change or variable that would exist between the control and your other test experiments would be a spot remover used in place of the water. All other variables would remain the same—for example, same size spot, same amount of spot remover, same method of applying the remover, same type of cloth.

As you conduct your experiments, record all your data in a journal. Take detailed notes of your materials, measurements, observations, and thoughts. Use headings and skip lines in between subjects. Every time you enter data or information in your log, note the date, the time, and any other important information. Include as much information as possible. Sometimes you may not realize how vital a detail is until after you have done several test experiments and reviewed your notes.

You may want to take photographs or make drawings and diagrams, especially if your project involves visual descriptions such as changes in color or patterns of behavior. Also take note of any problems you have. These observations will be valuable when it comes time to repeat—and possibly revise—your experiment and draw a conclusion. In addition, the judges may decide to ask you specific questions about what happened.

Keep in mind that you will have to repeat your experiments to make certain they support your conclusion. Because something happens once does not mean it will happen every time. Judges look for consistency in your observations and results.

After you have collected all your data, do any calculations that will help you draw a conclusion. For example, compute percentages and averages and put data into charts and graphs. These formats will help you review your data clearly. They may also be used in your science fair display.

Making sense of the results

Once you think you have enough data, you are ready to draw a conclusion. Do the data support or reject your hypothesis? If so, explain why and tell the value of this information. For example, "I conclude that the best spot removers do have chemical x. I conclude this because the data show.... Knowing this information, consumers can spend their money more wisely." Your conclusion must come directly and only from your data. If you cannot come to a conclusion, you may need to perform more experiments and gather more data.

The results were the same for each plant. In each experiment, the plant grew the same amount, even in the control experiment. From this data, I conclude that my hypothesis was not correct; my homemade fertilizer does not help these plants grow bigger and better.

Evaluate! If your hypothesis is not correct, what could be the answer to your question? Summarize any problems and successes you had doing the experiments. Do you need to repeat it? What would you do differently next time? List anything you learned about your topic or scientific research in general. The judges will be eager to hear this information.

Do not panic if your data fail to support your hypothesis. Proving a hypothesis wrong is a valuable step in scientific discoveries. You can propose a different hypothesis and a new set of experiments to test it. If time allows, do the new experiments. Offering this kind of information will go a long way in impressing the judges.

Pulling it all together

You are now ready to pull all your work together in the form of a written report and display for the science fair.

Preparing your report

Your report should summarize what happened in your experiments and explain your conclusion. Double-check your grammar and spelling. Have someone else read your report and offer you suggestions or advice.

Reports are generally typed double-spaced and bound in a folder or notebook.

Most science fairs require specific information be included with your report. Follow all directions carefully to ensure that your report is complete.

Designing your display

Your science fair display should summarize your project. Most displays include a three-

---

When putting together your report, use the following checklist as a guide.

__ An abstract, or summary, of the project. In one page or less, preferably three paragraphs, describe the purpose, procedure, and conclusion.

__ Any forms required by fair officials.

__ A title page. This is normally the only page with your name on it. Include your school and grade, too.

__ A table of contents. Double-check actual page numbers against the table of contents.

__ Acknowledgments. Although the thinking and technical work for your project is done by you, others may offer support, guidance, or equipment. Do not list family members by name; keep your anonymity for the judging.

__ Your purpose and hypothesis in two or three sentences.

__ A review of relevant literature that you read before and during your experimentation. This should be about two pages long.

__ A list of materials and equipment.

__ A description of experiment procedures. Explain what you did, all variables and controls, and how you measured and recorded what happened. This should be like a detailed recipe.

__ Your data. Number and label every chart and graph clearly and provide written summaries of them. Do not include conclusions, just data.

__ Your conclusion, based on your results. What is the value of the information?

__ A bibliography of all the sources you used.

__ Your notebook with all notes and observations, just as you took them— this item should not be retyped or rewritten.

__ Re-read your science fair rules and double check that you have all the necessary items included in your report.

sectioned, free-standing backboard, your report, some of the materials used in testing, and maybe a demonstration. But not all displays are the same, and sometimes being different can help your project stand out. Select the most important visuals from your project. Use photographs to show things that can't be displayed at the fair. But avoid using too many, or your display will look crowded or busy. Keep it simple.

Neatness counts. Consider using a computer to make headings and graphs. Clearly label each visual. A ruler will help you place things in a straight line. Plan for holes in your display for electrical cords if necessary. Instead of having tape or staples showing, use rubber cement to stick things to your backboard. Don't display hazardous items, such as needles, knives, and scissors.

Before constructing your display, draw a layout to help you plan where to place all the items. Your display must include a title; a summary of the problem, or question; your hypothesis; procedures and materials used; data, or results; and conclusion. If anything is missing, the judges will notice right away.

A Typical Display

The three-sectioned background can be made of wood, strong poster board, cardboard, or plastic, but check your rules—some fairs have fire safety guidelines that do not allow cardboard.

The title of the project should be large, neat, and spelled correctly.

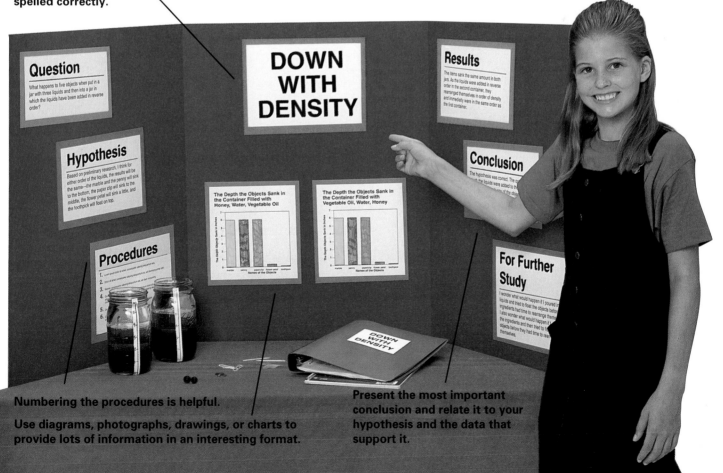

Numbering the procedures is helpful.

Use diagrams, photographs, drawings, or charts to provide lots of information in an interesting format.

Present the most important conclusion and relate it to your hypothesis and the data that support it.

Showing off yourself and your work

With all your hard work behind you, it's time to show off your project, share your findings, and have some fun.

Presenting to judges and others

Interviews are an important part of science fairs. The judges will ask you questions to assess your understanding of your project. The judges aren't there to challenge you. Often, they will offer valuable feedback and insight on your project. Review the list on this page. Once you know what the judges are looking for, you can better prepare for the science fair and anticipate the types of questions they will ask you.

In the days before the fair, review your procedures, research, data, and conclusions. Know your project thoroughly. Rehearse what you will say and

---

Know the five main judging criteria.

__ **Creativity** Is your problem an original one? Were you creative and original in solving the problem, designing the experiment, and analyzing and interpreting the data?

__ **Scientific thought or engineering goals** Did you use the scientific method? Was your problem too easy or too complex for your level? Did you clearly define your problem and procedures? Did you recognize all variables and controls? Did you gather enough data? Did you recognize the limits of the data and relate them to other research in the field? What further research could be done? If your project is engineering-based, did you have a clear objective? Is your solution feasible? Did you test the solution under realistic conditions?

__ **Thoroughness** How detailed are your notes? Did you track changes in procedures? How much time did you spend on the project? How complete was your research? Are you aware of other literature in the field and other approaches to the problem? Did you run enough tests?

__ **Skill** What resources did you have available to you? What help did you have?

__ **Clarity** Do you understand your project? Can you explain it to others? How clear are your written materials and your display?

how you will present material. Record your presentation on audio or videotape so you can review it and make changes as necessary. Explain your project to parents, teachers, and friends.

Handle all questions with poise and politeness. Listen carefully to each question. Take a moment to gather your thoughts before answering. Never be afraid to say you don't know the answer or that something did not come up in your research. It is better to admit to not knowing the answer than to pass along false information.

Speak slowly, clearly, and correctly, particularly when discussing technical information. Remember, nobody knows your project as well as you do. Avoid saying "like," "uh," "you know," "sort of," and "um" between words and phrases. Just pause instead.

Project a positive self-image. Be neat and clean and avoid wearing distracting jewelry or clothing. Remember, no eating, drinking, or chewing gum.

Use positive body language to convey your confidence and enthusiasm. Stand up straight, but not stiff. Maintain eye contact. Use gestures or props from your display to add emphasis or to clarify an important point. Smile. If you are enthusiastic about your project, others will be too. Shake each judge's hand and say thank you at the end of your discussion.

Getting the most out of the fair

No matter how well prepared you are, you will probably feel nervous on the day of the fair. To help relieve the tension, stretch out tense muscles. Jump up and down a few times to release some energy. Breathe slowly and deeply. If you look confident, no one will know if your stomach is in knots. The other presenters are anxious, too. Remember, you're at the science fair to meet people, learn some new things, and share your work. Take pride in yourself that you completed the project and presented it at the fair.

Space

Astronomers study the skies to answer questions about the universe: how big it is, what it contains, and how it changes. Earth's solar system is only a minor member of the Milky Way galaxy. This galaxy is about 100,000 light-years wide and is one of about 100 billion galaxies in the universe.

This seemingly endless space holds many mysteries. From living in a space station to interpreting amazing photos from space, scientists find the universe an exciting laboratory.

In this chapter, unravel some of the mysteries of space by exploring *Solar power*, *Finding stars*, and *Living in space*. Each project is a starting point for answering basic questions about space. Some of these questions follow. Can you think of others? Each one could lead to an exciting science fair project.

Finding stars

How do pollution levels affect our ability to view stars?

What conditions provide the best viewing of stars in your area?

How does the viewing of stars differ from a city to the country?

What is the best method for locating a star in the sky?

Solar power

How can solar energy be harnessed to help heat and light a room?

Can solar energy be used to cook food? How could it be done?

How do different clothing materials absorb heat from the sun?

How does the angle or shape of the reflector affect the amount of heat the solar collector reflects onto a ball?

Living in space

How is a bottle garden similar to or different from Biosphere 2 or a space colony?

By studying a bottle garden, can you identify components that are critical to the health of its inhabitants? Is it possible to isolate the variables that determine success?

Do the size of the bottle and number of plants affect success or sustainability? How?

What else could be tested as a part of a bottle garden? Does adding worms or insects improve conditions?

18 Solar power

On a sunny day you can feel the heat coming off the ground, once the **energy** of the sun has heated it. The sun supplies enough energy to each square yard of Earth's surface to power several electric light bulbs. If we could trap just a fraction of this energy, we would no longer need to burn polluting fuels, such as coal and oil.

You will need

tape, cardboard, scissors
large aluminum foil
 pie pans
two thermometers
an awl, black paint
wooden dowels
ping-pong balls
clay flowerpots

1 Make a hole in the center of the pan. Cut from the edge of the pan to the hole. Overlap the cut edges so that the base of the pan curves like a satellite dish.

2 Paint a ball black. Ask an adult to use the awl to make two holes in it at a 90° angle to each other.

3 Use cardboard and tape to fix the pan at an angle on top of the flowerpot.

Discover!

This simple **solar** heater uses a curved reflector to concentrate sunlight and heat the air inside a ball. Solar power stations in the United States and elsewhere use the same method. Large mirrors focus sunlight on water pipes or boilers. Steam from the boilers turns turbines to make electricity.

4 Push the dowel through the hole in the pan. Mount the ball on top of the dowel. Place your solar heater in the sun. Adjust the angle of the pan and the height of the ball until sunlight is reflected onto the ball.

Now you are ready to test your solar heater. Use the thermometers to compare the temperature of the surrounding air to the temperature inside the ball. Because Earth is constantly spinning, you will need to move the reflector dish to keep the sun focused on the ball. In real solar power stations, the mirrors are moved automatically by computer, so that they track the sun as it crosses the sky.

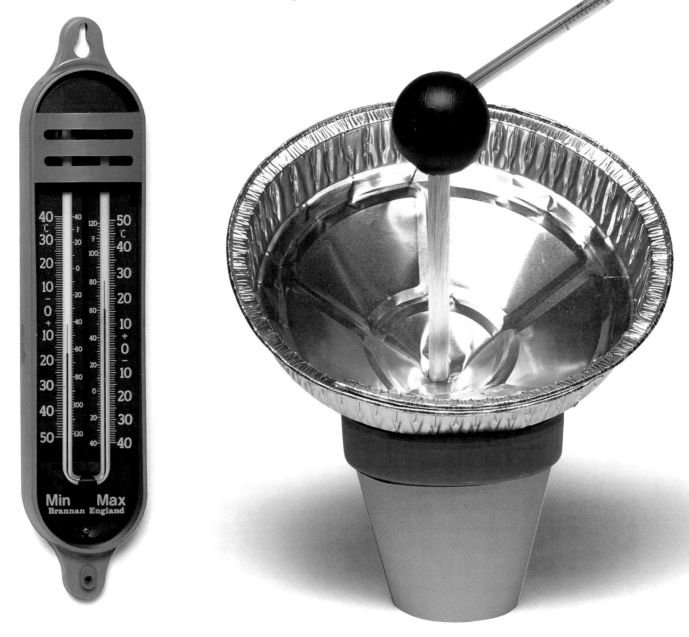

Experimenting with your solar heater

Make a note of the temperature inside the black ball after a few minutes. Record the temperature every hour for several hours. Create a solar heater with a white ping-pong ball and repeat the experiment. How do you think the results will compare? When are the temperatures highest? Lowest? Note the temperature at the same time of day every day for a month. Note the weather conditions at the time of your readings. How does weather affect the amount of the sun's energy absorbed by the ball? If you could keep records over several seasons, how do you think the temperature would change in summer? In winter? Would your results be the same if you used a ball painted a different color? Colored cloth instead of a ball? What if you didn't change the angle of the pan?

The space pioneers who may eventually set off to colonize Mars will have to create their own sealed **environment** in which to live. Growing plants to provide enough food and oxygen will be their main priority. The pioneers will use solar energy and will have to recycle all their water and waste materials.

During the day, the plants' green leaves collect energy from sunlight, which they use to convert water from the soil and **carbon dioxide** from the air into a simple sugar called glucose. This process is called **photosynthesis**. As the leaves make glucose, they also make oxygen. Plants and the tiny animals (worms or insects) in the bottle use the food (glucose) and oxygen to grow. When they do this they release water and carbon dioxide. This is called respiration.

In a bottle garden, the sun's energy constantly recycles oxygen, water, and carbon dioxide between the soil, the living things, and the air.

You will need
water
a trowel
potting soil
extra-large glass jars with lids
plants (Small, well-established plants that like a warm, damp atmosphere, such as ferns, miniature ivy, African violets, mosses, begonias, and small spider plants, are the best.)

Discover!
A space colony could work in a similar way to a bottle garden. Plants sealed inside the bottle use the energy of sunlight to grow.

1 Fill the bottom of the jar with a layer of potting soil, about 4 to 6 in. (10 to 15 cm) deep. Level it off and pat the soil firm.

2 Gently remove your plants from their pots. Using the trowel, make holes in the soil large enough to accommodate the roots of the plants. Put the plants in the holes and replace the soil, pressing it down firmly.

3 Water the garden carefully. The potting soil should be moist but not too wet.

4 Put the lid on the bottle and seal it tightly. Place it where it will receive plenty of sunlight but not extreme, direct sunlight. The plants should grow without extra water. If the inside of the bottle clouds up completely, take off the lid for an hour or so to let it clear.

Experimenting with your bottle garden
How is your bottle garden like a space colony? How is it different? Which elements are critical to the success of a self-contained environment like a space colony? What tests can you think of? Create another bottle garden to use as a control. Is a bottle garden more successful with worms and insects, or without? Why? Is it possible to monitor temperature or humidity? Do they change? How do they affect the success of a bottle garden?

Biosphere 2 is an experimental environment in Arizona that has been run somewhat like a space colony. Six people lived inside the glass and steel dome for two years to see if it was possible to grow all their own food and recycle their water and air, *sealed off from the outside world. They lost weight and had some problems with the atmosphere, but they survived. More experiments like this will be essential before real space colonies can be built.*

As Earth spins, the stars seem to move in the sky. Earth's night side faces in different directions at different times of the year, so some stars that can be seen in winter are not visible in summer.

1 Cut out two circles from cardboard: one should be about an inch smaller in diameter than the other.

2 Buy or borrow a star chart. Make a photo copy of the the stars that are visible from the Northern Hemisphere. Glue the copy onto your larger disk. Trace over the constellations with luminous paint.

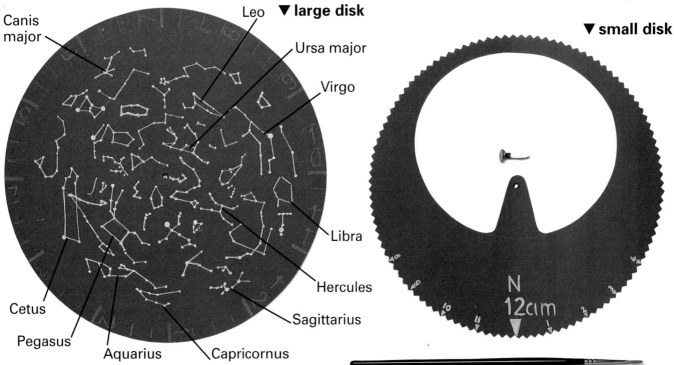

▼ **large disk**

Canis major
Leo
Ursa major
Virgo
Libra
Hercules
Sagittarius
Capricornus
Aquarius
Pegasus
Cetus

▼ **small disk**

Finding constellations in the night sky can be difficult. A planisphere is a special chart that shows where constellations are at different times of the year.

Discover!
You can make a model planisphere to find out how it works. If you want a more accurate planisphere, you must buy one specially made for the **latitude** where you live. The star finder shown on this page works for northern Europe and the northern parts of North America.

You will need
cardboard	glue
scissors	a star chart
luminous paints	a brad
paintbrush	a magnetic compass

3 Make 12 equally spaced marks around the edge of the large disk. These represent the months. Using your paints, number the months from 1 to 12 and divide each section into four as shown above left.

4 Cut a window from the small disk, as shown above. Leave a "finger" of cardboard reaching to the center.

5 Mark the hours from 8 P.M. to 4 A.M. on the edge of the small disk. The space between each hour should be half the space given to each month on the big disk. Mark north with an N and an arrow at 12 midnight as shown.

6 Make a hole in the center of both disks and join them with the brad.

To use your star finder, turn the upper disk to line up the time with the month. Now, use a compass to find magnetic North and turn to face it, holding your star finder in front of you. If the night sky is clear, you should be able to see the stars shown in the window.

▲ The night sky at midnight, January 1

With your naked eye you can see about 3,000 stars, but with a pair of binoculars you can see many more. All the stars we can see are part of a huge "star city," or **galaxy,** *called the* **Milky Way.** *There may be as many as hundreds of billions of stars in the Milky Way, but it is not the only galaxy in the universe. There are billions of galaxies, each made from gas, dust, and billions of stars, many of which are like our sun. Most are so far away that they can only be seen with the use of a powerful telescope. The light from the farthest galaxies takes billions of years to reach Earth.*

Try to view the night sky in an area where there are no street lights, otherwise you will only be able to see the brightest stars.

Experimenting with your planisphere
The stars seem to sweep across the sky each night. Also, the positions of stars in the sky seem to change slightly over many years. What does this motion tell you about Earth's

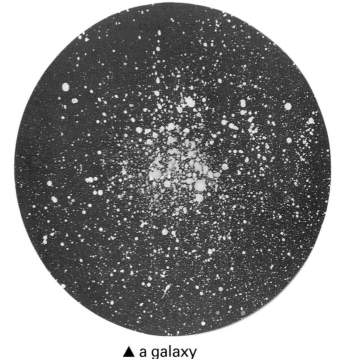

▲ a galaxy

movement? About the solar system? People have charted the positions of the stars for thousands of years. Why was this important? If you buy a planisphere for a different latitude, is the window in a different position on the upper disk? Compare different methods for finding stars. Which works best?

People still watch the sky, and astronomy is one of the few sciences in which amateurs can make valuable contributions. Astronomical societies provide information on astronomy for their members by publishing newsletters and holding meetings. Write to an astronomical society; find out what other astronomers are looking for. What discoveries might you make?

Dynamic Earth

Humid rain forests, arid deserts, ice-covered peaks, islands that are the tips of mountains looming from miles below the sea—these are just a few contrasts found on Earth's surface.

Earth's most striking feature, however, may be its dynamic nature. As Earth spins on its axis, the surface moves at over 1,000 miles (1,609 km) an hour, helping to produce wind and weather (but luckily not motion sickness!). The huge plates we live on shift a few inches over one year. Rocks and minerals are constantly changing from temperatures of about 1600 °F (870 °C) deep in the crust and from erosion on the surface.

Explore Earth's physical features and weather as geologists and meteorologists do. Look into *Wind direction, Shifting plates, Rocks,* and *Soil* on the pages that follow. Each project is a starting point for answering basic questions about Earth. Some of these questions follow. Can you think of others? Each one could lead to an exciting science fair project.

Wind direction

Does wind travel in any particular direction in your area?

Is there any relationship between wind direction and weather? What about wind direction and time of day?

How does wind direction differ from one season to another? Why?

Does wind direction differ with the altitude of a location?

Shifting plates

Is there any evidence of plate movement around your community? What could be the cause of specific shifting around your house? Does it relate to the causes of Earth's crust shifting?

Can you simulate Earth's plate movements for the years to come? How might those movements affect the location of the continents?

Rocks

What minerals exist in your water at home? Do they affect your pipes and other plumbing fixtures? How?

What rocks or minerals make the strongest building materials? How can you test their strength?

What types of rocks exist in your neighborhood or region? What can you tell about the history of the area based on your rock findings?

Soil

What type of soil is the best for the plants in your region?

How do fertilizers affect the condition of soil?

How do small animals, such as worms and pill bugs, affect the condition of soil?

Which soils hold fertilizer the longest? (Or, in which soils do fertilizers wash away fastest?)

Winds are described by the direction from which they are blowing. Westerly winds come from the west, for example. Wind direction has a strong influence on the weather. In Europe and North America, for instance, northerly winds

usually mean cold weather. In the same way, if a wind is blowing across a stretch of ocean, it is more likely to bring rain than if it comes from across a desert or over a mountain range. In order to predict the weather, meteorologists study the wind direction over long periods.

Discover!
Make your own wind recorder and use it to build a picture of long-term wind patterns.

You will need
two empty thread spools
a shoebox, sand or stones
cardboard, colored paper
glue, a sharp pencil
wooden dowel
graph paper, tape
a compass

1 Glue an empty spool inside the bottom of a shoebox, slightly off-center. Weigh down the box with sand and stones.

2 Make a hole in the top of the box, above the spool. As shown, glue a piece of paper with a hole in the middle and slots cut in the corners to the top of the box. Then, glue the compass to the box as shown.

3 Use graph paper to make record cards. Put a hole in the middle and label the edges north, south, east, and west. Slot one card in the paper on the box top. Tape down the edges on particularly windy days.

Experimenting with a wind recorder
Slip a new record card onto the wind recorder each day and write down the daily readings in a notebook. Be sure to note other conditions as well such as date, time of day, temperature and cloud cover. Later, compile your data for analysis. From which direction does the wind blow most often?

4 Put the second spool over the hole in the box top. Pass the dowel through both spools. Check that the dowel can swivel freely.

5 Make a triangular wind pointer out of the cardboard and colored paper. Fix it to the dowel as shown. Tape a pencil to the side with the point touching the top of the box.

6 Set out your recording device. Put it where it will be protected from rain, but not from wind. The thickness of the pencil line will show where the wind has been blowing from most.

This will be the **prevailing wind** in your area. Does it change from season to season? How does wind direction correspond to temperature and weather? Is there any change on cloudy days? At higher or lower altitudes? Why do you think this is?

28 Shifting plates

We don't usually notice it, but the surface of Earth is moving all the time. The **tectonic plates** that fit together to form Earth's crust shift a small amount every year, about 4 inches (10 cm)—that's about as rapidly as human hair grows. Yet the effects of these tiny movements can be enormous. Where the edge of one plate slides beneath another, violent earthquakes may occur, and the pushing and pulling of plates over millions of years has created many of Earth's mountain ranges.

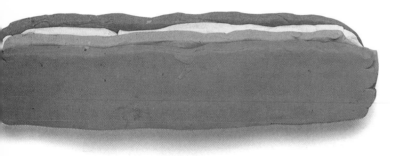

Discover!
The movement of Earth's tectonic plates has caused two main types of mountain ranges. In fold mountains, the crust of the Earth has wrinkled up into wavy folds. In fault-block mountains, the pressure of the moving plates has cracked brittle layers of rock, shifting and tilting blocks of Earth's crust. Make some simple models of faulting and folding, using strips of modeling clay.

To form mountains, you will need
modeling clay   a modeling clay cutter

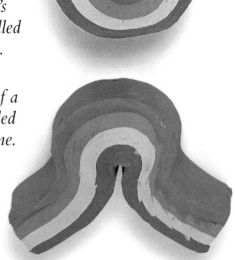

The bottom of a fold in the Earth's crust is called a syncline.

The top of a fold is called an anticline.

1 Take four or five strips of clay and lay them on top of one another, as though they were layers of rock in Earth's crust.

2 Hold the clay at both ends and push it together. It will bend into the shape of a fold mountain—either a syncline or an anticline.

3 Now take another strip of clay and slice through it once or twice with the clay cutter. Slide the pieces of clay against one another as though the rock were pushing up or down.

When faulting pushes a section of Earth's crust upward, the raised rock is called a horst. When the faulting thrusts a section of crust downward, it makes a rift valley.

Discover!

Make a jigsaw puzzle of Earth's plates. Use the puzzle to demonstrate how the plates move.

To make a tectonic puzzle, you will need

a world map, pencils
tracing paper, a red pen

thick cardboard
a utility knife, glue

1 Using your map, copy the world's major land masses onto tracing paper with a red pen. Then turn the paper over and rub the tip of a pencil along the back of the red line.

2 Turn the tracing paper over and put it onto the cardboard. Draw over the outline of the world map again, pressing down firmly. The pencil outlines will transfer onto the board.

3 Have an adult help you cut out the land shapes with the utility knife. Be careful! Cut away from you and hold the cardboard firmly.

4 Glue the land shapes onto another piece of cardboard. Copy the outlines of the plates onto your cardboard map. Cut out the plates.

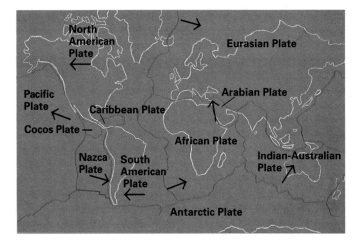

The boundary between two plates that are moving apart is called a divergent boundary. Where two plates are colliding, there is a convergent boundary. A transform boundary occurs where two plates slide along each other. Refer to the arrows on the map to label your puzzle. Research land formations that occur where plates meet. Can you draw any conclusions or make any predictions about these formations? Is there any evidence of plate movement around your community or in your state?

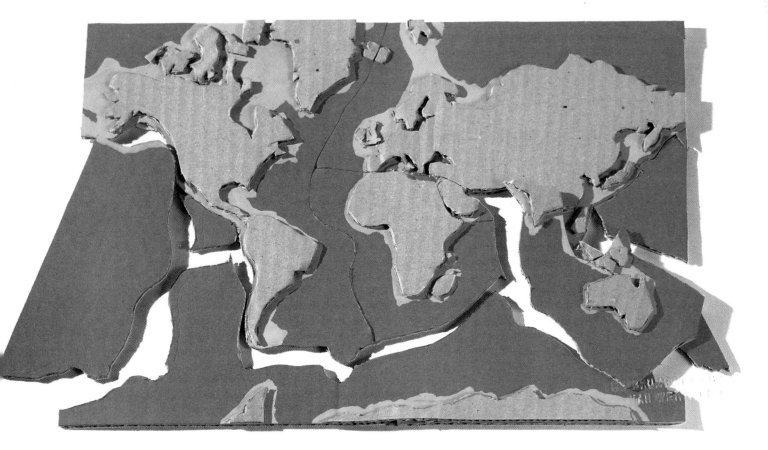

Rocks are solid lumps of **minerals** that make up Earth's crust. There are basically three different ways in which rocks are created—so rocks are classified by geologists in one of these three categories:

● Sedimentary rocks. They are made of layers of sand and silt that have been squashed together.

● Igneous rocks. They are rocks that became so hot that they melted but then cooled down again and solidified.

● Metamorphic rocks. These are sedimentary or igneous rocks that have been altered by heat or pressure.

Discover!

Geologists study rocks carefully, examining their texture and the way they have been formed. Make your own collection. Label them, classify them, and draw conclusions.

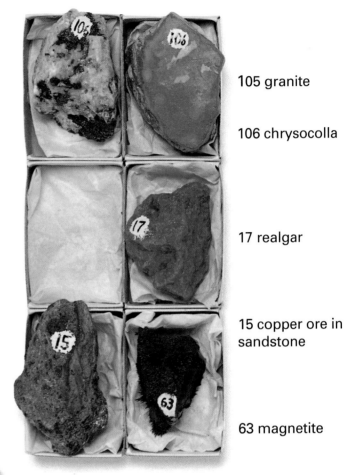

105 granite

106 chrysocolla

17 realgar

15 copper ore in sandstone

63 magnetite

1 Collect some samples of the rocks in your area. Make some specimen holders out of empty matchboxes for your collection of rocks. Inside each compartment put a layer of tissue paper for the rock to rest on.

To start a rock collection, you will need

a geologist's hammer	tissue paper
a magnifying glass	correction fluid
plastic bags	a notebook and pen
empty matchboxes	a reference book on
glue	rocks and minerals

2 Number the rocks. Dab white correction fluid in one corner and write the number in ink.

3 Make a fact sheet for each rock. Draw the rock and note where it came from. If you don't know what it is, look for it in a guidebook.

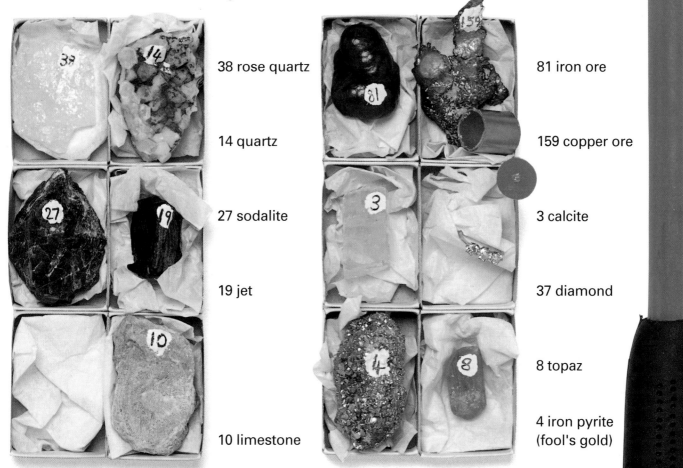

38 rose quartz

14 quartz

27 sodalite

19 jet

10 limestone

81 iron ore

159 copper ore

3 calcite

37 diamond

8 topaz

4 iron pyrite (fool's gold)

All minerals have specific characteristics, such as chemical composition, hardness, and color. What tests can you do to determine each rock's characteristics? Minerals can also be identified by noting the formations in which they are found.

Many beginners purchase a reference collection of rocks from a rock and mineral dealer. These collections identify common rocks. You can identify unknown rocks by comparing them with pictures and known specimens.

What do your findings tell you about the geologic history of your area? Also, how do you think the presence of these minerals affects the water supply? The soil?

Discover!

Limestone contains the mineral **calcite**. Spectacular formations grow in limestone caves. As water passes through a cave, it dissolves calcite from the surrounding rocks. Then, as the water drips and dries, it leaves behind deposits of calcite that gradually build up into **stalactites** and **stalagmites**. Make your own stalactites and stalagmites and experiment with them.

Hard and soft rocks

A scale of hardness for minerals was developed by the Austrian scientist Friedrich Mohs in 1812. The scale, based on the rocks shown below, is from 1 to 10. At one end of the scale is the diamond, which can scratch anything. At the other end is talc, which is the softest.

To make stalactites, you will need

glasses   string
distilled water saucers
soda crystals   paper clips

1 Half fill both glasses with distilled water. Gradually pour in as many soda crystals as the water can dissolve. This mixture is called a **saturated solution** of soda.

2 Dip a piece of string into the solution. Run it from one glass to the other, with the saucer in between. Fix the string in place with paper clips and wait a few days.

▼ *Mohs based his hardness scale on the ten types of rock shown below.*

1 talc

2 gypsum

3 calcite

4 fluorite

5 apatite

6 orthoclase

7 quartz

8 topaz

9 corundum

10 diamond

3 The solution travels along the string. When it reaches the lowest point, the solution drips into the saucer. Water **evaporates,** leaving soda deposits to form a stalactite. A stalagmite will also grow upward from where the solution drips onto the saucer.

Experimenting with stalactites/stalagmites
Create a control and then experiment with one variable at a time. What conditions produce bigger formations? Can you speed up or slow down the process? How do these tests relate to conditions in a cave?

Discover!
Sugar crystals are not minerals, rather, they are organic (formed from living matter). However, sugar crystals can be used to show the ways that mineral crystals form. Make these crystals and experiment with them.

To grow crystals, you will need
distilled water, table sugar
cotton thread
paper clips, pencils
1-pint (500-ml) jars
a small saucepan

1 Pour ½ cup (or 120 ml) distilled water into the saucepan. Ask an adult to help you heat the water over medium heat. Slowly add 1¼ cup (or 300 ml) sugar, stirring continuously until the sugar is dissolved.

▲ *Stalactites with a **c** grow from the ceiling. Stalagmites with a **g** grow from the ground.*

2 Remove the pan from the heat once the liquid starts to boil. Allow the solution to cool for 15 minutes. Then, pour the solution into a jar.

3 Tie one end of a piece of string to a paper clip, and the other end to a pencil. Lower the paper clip into the jar of solution until the paper clip rests on the bottom of the jar. Lay the pencil across the opening of the jar.

4 Observe the jar's contents for at least two weeks. Make diagrams and keep notes.

Experimenting with crystals
Can you make crystals grow faster or larger? Create a control and then test one variable at a time. What happens if you use twice as much sugar? Half as much? What if you don't heat the water? What if you use ice water? What happens if you don't use distilled water? Which of the above variables do you predict will or will not affect crystal formation? Why? How do these tests relate to conditions in a cave?

Soil is made from rocks and minerals that have been ground up and mixed with water, air, and the remains of dead plants and animals. As with rocks, there are different types of soils. Some soils are better for growing crops than others.

Discover!
Soil settles in different layers, or **strata**. The largest, heaviest particles are usually those at the bottom (the **subsoil**), with the finest particles in the **topsoil** nearest the surface. Analyze your own cross section of soil layers.

To analyze soil strata, you will need
a plastic bottle	water
a mixture of soils, dug out of one hole in the earth at different depths	plastic tubing
	a thick needle
	a craft knife
	cardboard, tape
poster board	a jar or glass

1 Using a funnel made from poster board, half fill the bottle with soil and top it off with water. Put the top on the bottle and shake it well. Then let it stand for several hours.

Mix soil from one strata with a ▶ little water and smear the paste onto white poster board. Do this for each strata of soil. Compare the smears. What do you notice about the colors?

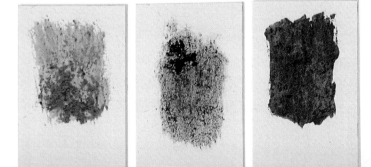

2 Siphon the water, taking great care not to disturb the settled soil. Put one end of the plastic tube into the bottle. Suck a little on the other end. As soon as the water starts to move in the tube, quickly take the tube out of your mouth and put it into a jar that is on a lower surface. Once it gets started, the water will keep flowing into the jar.

3 To drain off any last drops of water, very carefully make small holes with a thick needle in the bottom of the bottle. Let the bottle stand for a few minutes in a sink.

4 Ask an adult to cut off the top of the bottle. Tape cardboard over the opened area. Gently lay the bottle on its side. Carefully slit it lengthwise in two. Slide a piece of cardboard down the slit as you cut so that no soil falls out.

Discover!

Soils may contain chemicals that are either **acids** or **alkalis**. Scientists measure these chemical properties on something called a pH scale. pH generally ranges from 0 to 14. A pH of 7 is neutral. A pH below 7 is acidic, above 7 is basic (alkaline). Acidic soil is less likely than alkaline soil to be rich in the minerals that crops need. Perform your own soil test. Use litmus paper, which reacts to acids and alkalis.

To test soil acidity, you will need

glasses   distilled water
soil samples   litmus paper strips

1 Put soil in a glass and add distilled water. Mix well. Then let the glass stand a few minutes.

▲ *Acids turn purple litmus paper pink.*

▼ *Alkalis turn pink litmus paper purple.*

Experimenting with soil samples

Analyze and identify the strata in your soil samples. Take soil samples from different areas—from a garden, playground, backyard, farmland, or riverbank. Which do you think is the most fertile? Why? How do fertilizers affect the soil? Do they stay in certain soils longer? How do small animals, such as worms and insects, affect the soil? How do people affect soil, both negatively and positively? What happens to nutrients in the soil when insecticides are used? Are there other tests you can do to find out more about your samples?

2 Dip a pink litmus strip into the water above the soil and see what happens. If the soil is acid, the color will not change. If your soil is alkaline, the pink paper will turn purple.

3 Now dip a purple litmus strip into the water and see what happens. If your soil is acid, the purple strip will turn pink. If it's alkaline, the color will not change.

4 If neither the pink nor the purple litmus paper changes color much, you should classify your soil as neutral—it is neither very acid nor very alkaline.

Machines and Structures

People use machines to build sky-scrapers, lift heavy loads, and move over the earth, through the air, and under the sea. People even make machines that can travel through space.

The scientists who build such machines and structures are called engineers. They do tests and experiments that help them invent new machines and make old ones work better. They design tools, engines, airplanes, ships, towers, bridges, and more.

The engineer's job is not easy. The key to the science of engineering is in understanding mathematics, physics, and chemistry; choosing the right materials; and knowing how to put them together correctly.

The projects in this section—*Submarines, Airfoils, Boats, Bridges,* and *Power of shapes*—help develop your engineering know-how. Each project is a starting point for answering basic engineering questions. Some of these questions follow. Can you think of others? Each one could lead to an exciting science fair project.

Submarines

How does a submarine dive and resurface?

Is there another way for a submarine to dive and resurface? Is there a more efficient way?

What factors affect buoyancy?

Bridges

Can you design a strong but lightweight bridge?

What makes a bridge better able to withstand an earthquake?

Is an arched bridge stronger than a flat one? Why or why not?

Is a suspension bridge stronger than a nonsuspension bridge?

Boats

What limits the speed of a boat? Why?

What designs make boats move faster? Straighter?

Which design is most stable?

Which kind of boat can carry the heaviest load?

Airfoils

What is lift?

What are the effects of size and shape on lift?

What kite design has the best lift?

What is the most efficient airplane design?

Power of shapes

Which is stronger, a structure shaped like a triangle or one shaped like a square? Why?

Can a strong chair be made without nails, screws, or glue?

What are tension and compression? How do they influence structures such as chairs, bicycles, and skyscrapers?

A submarine dives by flooding tanks called **ballast** tanks with water. The added weight causes the ship to lose its ability to stay afloat. Then horizontal rudders called hydroplanes are tilted down, and the craft glides down into the water. To resurface, water is blown out of the ballast tanks by compressed air, or the hydroplanes are tilted so the submarine angles up.

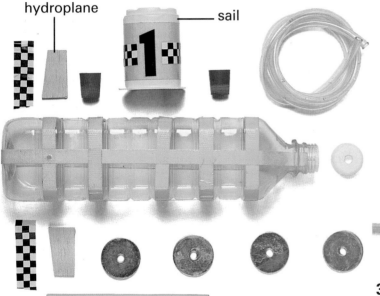

1 Trace the bottom of a bottle stopper on each end of one side of the plastic bottle. Ask an adult to help you cut out the traced shapes, using a craft knife. Fit the bottle stoppers tightly into the holes, as shown below. Now cut two slightly smaller holes in the opposite side of the submarine.

2 On the bottle's sides, make two small holes across from each other toward the back of the bottle. Push a short dowel in one hole, through the bottle, and out the other hole. Cut two balsa-wood hydroplanes in the shape shown. Glue a hydroplane to each end of the dowel as pictured below.

3 Ask an adult to help cut a hole in the bottle cap that is the same size as the tubing. Push the tube into the cap, seal the hole around the tube with clay, and screw the cap on tightly.

4 For the ship's sail, decorate a yogurt container. Make a hole in the bottom and push a dowel through it to make a periscope. Make another small hole in the side of the container so it can fill with water. Glue the container upside down to the submarine.

5 Tape several washers or coins to the bottom of the submarine at the front and back. Tape enough to help the submarine sink when full of water, but not so many as to make it sink when full of air.

Discover!

Can you build an efficient submarine? Like a real submarine, this model dives and resurfaces by using water and air.

You will need

a craft knife	a hand drill
balsa wood	modeling clay
plastic tubing	thin dowels
large washers or coins	duct tape, glue
plastic bottles with caps	yogurt containers
rubber bottle stoppers	tub or waterproof tank

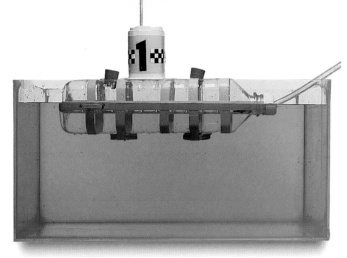

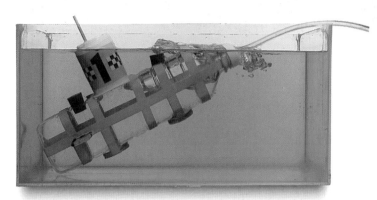

6 Put the submarine in water with the bottle stoppers in. When the submarine has taken in a little water from beneath and floats at the surface, stop more air from escaping out the tubing by pinching it.

8 To make the submarine rise, put the stoppers back in while it is still underwater. Now blow hard into the plastic tubing. As this air fills the tank and pushes water out through the bottom holes, the submarine becomes lighter and rises to the top of the tank.

9 When you have blown out nearly all the water, the submarine will sit on the surface of the water. As soon as you stop blowing, pinch the tubing to stop the air escaping or water will flood back in and the submarine will sink again.

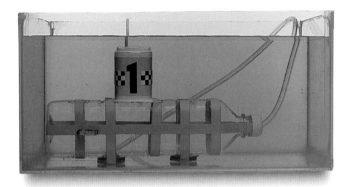

7 Next, take out the top two stoppers and release your grip on the tubing. Air will escape out of the tubing, allowing water to rush in through the holes in the bottom of the submarine. The vessel will sink.

Experimenting with your submarine

Can you make your submarine dive and resurface faster? Does the temperature of the water make a difference? How does the size or shape of the hydroplanes affect the efficiency of your submarine? Can you think of other factors to test? Be sure to set up a control.

If you were to look at the wings of an airplane, you would find that the top surface of the wing is curved, while the bottom surface is more flat. Most wings have this shape, called an airfoil. Air moves around the airfoil as the airplane flies. This provides the **lift** that keeps the airplane in the air.

Discover!

Use a fan to investigate the force that keeps an airplane in the air. To be extra safe, use a fan with a cage or rubber blades, like the one shown here.

1 Cut a rectangle of poster board 4 in. x 12 in. (or 10 cm x 30 cm). Draw a line across the board $2/3$ in. (1.5 cm) from one of the short edges.

2 Fold the poster board in half so that the opposite short edge touches the line. Glue $1/3$ in. (1 cm) of the edge down. You should have a shape with a curved top. This is an airfoil.

3 Ask an adult to help make two holes through the center of the airfoil's thickest part, one directly above the other.

4 Push the bead onto one end of the wire, bending the wire to hold the bead in place. Push the other end of the wire through the holes in the airfoil.

You will need

a protractor
cotton thread
a drill or an awl
a wooden block
a desk fan
poster board
stiff wire
scissors
a bead
glue
tape

5 Make a hole in the middle of the wooden block, just big enough to hold the end of the wire. Push the wire into the block so that it stands firmly in the wood. You should be able to change the angle of the wire.

6 Tape the protractor to one side of the block.

7 To help you see how air moves over a wing, tape pieces of thread to the airfoil.

Lift forces the airfoil upward along the length of the wire.

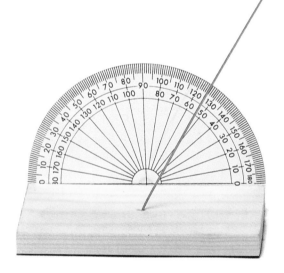

flexible flap

▼ *Use the protractor to measure the angle of the airfoil to the airflow.*

8 Stand the fan in front of the airfoil, making sure that the rounded edge is facing the fan. This edge is the leading edge, and the other is the trailing edge. The curve of the airfoil makes air moving over the wing travel faster than air moving below the wing.

Fast-moving air does not press against objects as much as slow-moving air. Therefore there is less pressure on the top of the wing than there is on the underside. The higher pressure on the underside pushes up and lifts the wing.

Experimenting with airfoils

Adjust the angle of the airfoil to the airflow by adjusting the wire. How does this affect lift? Which angle is the most efficient? Does the speed of the air flow affect lift? Try making airfoils out of different shapes, sizes, or weights. Can you relate your findings to shapes of helicopter propellers, airplane wings, kites, or other items?

When the airfoil is turned upside down, the air forces it to the ground. To produce an upward force, the curved side of the wing must be on top.

The airfoil creates lift with its flat edge at a small angle to the stream of air. It is pushed partway up the wire.

The size of the lift force increases with the angle of the airfoil to the wind. Here the airfoil travels quickly to the end of the wire and stays there.

The part of the boat that has the most effect on how the boat moves is the section in contact with the water: the **hull.** Design your own hulls and see for yourself. Make a chart of your findings. Check out docks, books, and magazines. Can you design a more efficient cargo ship or racing boat?

Discover!

There are two main hull shapes: broad hulls and long, thin hulls. See how different shapes travel by making some simple hulls and pushing them around in a tank of water. How quickly do they move? How stable are they if you swirl the water about? Use small weights to determine which is best for carrying cargo. Which is best for traveling fast? Why?

To test hull shapes, you will need

pieces of balsa wood   craft knife
plastic planting tray   modeling clay
tank full of water
small weights or pebbles

▼ *When a block of wood is put into the water, it may tip to one side. This is because the **density** of the wood can be uneven, making the block very unstable.*

1 Cut different shapes from balsa wood, including chunky blocks, and others like the long, thin one below. How quickly do they move through the water when given a push? Which take the most energy to move?

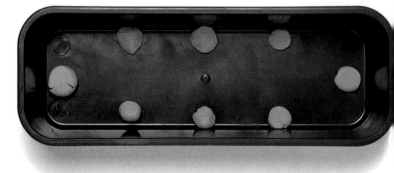

2 Take a plastic planting tray and make it watertight by filling in any holes with clay. Place the tray in a large tank of water. See how it floats, with its wide, shallow hull. How easily does it move through the water? How stable is it compared to the wooden ones you made?

Experimenting with hulls

Make hulls from different types of materials and of different shapes. Which would be best for use in shallow water? For long distances? Speed? Which do you think are the most stable? Consider the structures on real ships and boats. Is there a way to make a hull more stable or efficient? Why would this be helpful?

Discover!

The **keel** is the part of the hull that keeps the boat stable and stops it from being blown sideways or overturned by the wind. Compare broad keels with slim, narrow ones.

To compare keels, you will need

balsa wood, a craft knife a hacksaw, tape
glue, modeling clay a hand drill, pliers
thick wire a sheet of brass

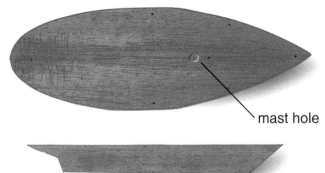

mast hole

1 Cut a wood hull 3 x 8 in. (7.5 x 20 cm), and a keel 9 in. (23 cm) long x 2$\frac{1}{2}$ in. (6.5 cm) deep. Drill a mast hole. Put clay along the keel for weight and stability. Glue the pieces together.

keel

Experimenting with keels

Float the large-keeled hull and the narrow-keeled hull in water. Set up tests for the boats. Which moves faster? Straighter? Which would

2 Cut the hull and cabin shapes from balsa wood. Drill a small hole at the back of the boat for the **rudder** and one at the front for the mast. Glue the cabin to the hull.

3 Ask an adult to help use the hacksaw to cut the brass keel. Use pliers to bend over one short edge at a right angle and glue the flap to the bottom of the hull.

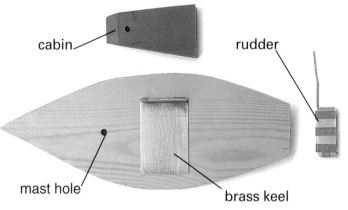

cabin

rudder

mast hole

brass keel

4   Ask an adult to help use the hacksaw to cut a brass strip for the rudder. Bend the wire into a long L-shape as wide as the rudder. Tape the rudder into the corner of the L. Push the other wire end through the hole at the stern. Bend it over to hold it in place and make a **tiller**.

tiller

hold up best in stormy weather? Research and consider the effects of **drag.** Consider adding sails to each boat and experiment with them. What other factor could you test? Remember, make a hypothesis and take notes.

Stone and brick are strong, but they are very heavy. For a long bridge, which has to support its own weight as well as the loads that cross it, lighter materials are needed. Engineers use steel girders, linked together in **lattice** patterns, to build long bridges that are strong, light, and not too expensive.

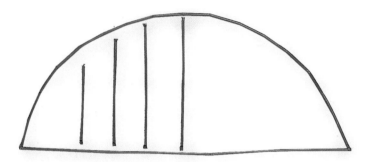

Discover!

Bowstring bridges are named after the string of an archer's bow. **Tension** in the bridge's bottom girder—the bowstring—holds the arch in place, just as tension in the string curves a bow. Find out how the **ties** carry the strength of the arch to the platform by building your own bridge.

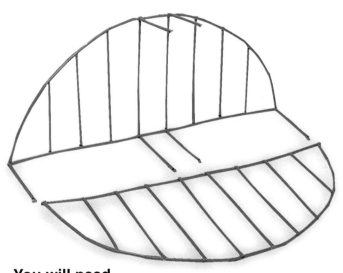

You will need

bricks	scissors, glue
plastic straws	thin cardboard
weights or yogurt cups and sand	

1 First, make the bow-shaped sides of the bridge using pieces of plastic straw. The piece that you bend to make the bridge's arch should be about one and a half times as long as the girder (the bowstring) that runs along the bottom.

2 You can join pieces of straw by putting a drop of glue on the end of one piece and pushing it into the end of another. You may have to split the end of one straw a little.

3 To attach the ends of the bowstring girder to the arch, split the ends of the bowstring straws. Open out the ends to make small flaps, then cut one of the flaps off each end. Glue the remaining flaps to the ends of the arch. Use this method to attach the ties between the arch and bowstring.

5 Cut a strip of thin cardboard to make the bridge deck and glue it in place. Your bridge is now complete.

6 Rest the ends of your bridge on two bricks. Test its strength by loading it with weights, such as yogurt cups filled with sand.

4 Join the two sides of the bridge with plastic straws of equal length. Make flaps in the ends of the crosspiece straws and glue them to the arches.

Experimenting with bridges

Look up pictures of various bridges and examine their designs. Research bridge building to see if you can explain why certain designs are so strong. Then, build several bridges with different designs, such as the ones pictured here. Or, use different amounts or kinds of materials. Predict which bridge will be both light and strong. Which bridge will be the strongest? Weigh each bridge, and test the strength of each bridge by putting weights on it. What other factors need to be considered when building a bridge? Wind? Rain? Heat?

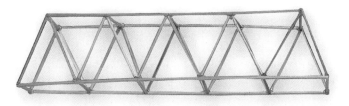

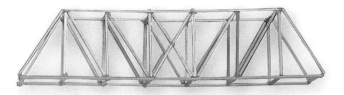

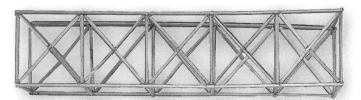

Some shapes are naturally stronger than others. Take a look at bridges, towers, desks, and chairs. What shapes do the structures' tubes and girders make?

1 Ask an adult to help you drill a hole about 1 in. (2.5 cm) from each end of each pole.

2 Fasten the poles together with the ropes as shown above. First tie a thick knot 4 in. (10 cm) from the end of one rope. Thread the long end of this rope through a hole in the first pole. Tie a second knot in the rope on the other side of the hole, so that the pole is held firmly between the two knots.

Discover!
Discover the strength of shapes. This stool is made from just three wooden poles and two pieces of rope. There are no nails or screws, yet it can hold the weight of a person. It gets its strength from its shape.

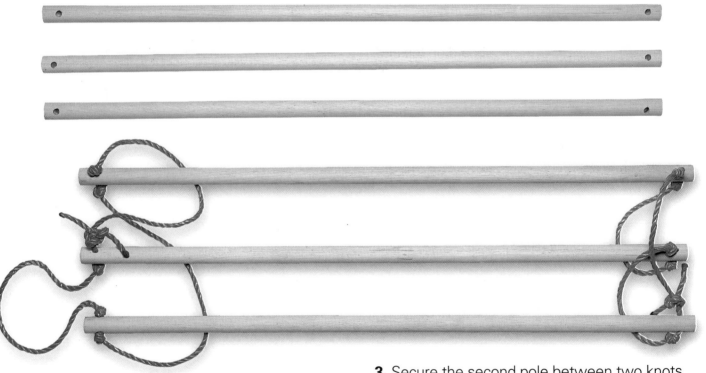

You will need
an awl	a large board
a vise and a drill	a tape measure

two pieces of strong nylon rope, about 5½ ft. (2 m) and 4½ ft. (1.5 m) long

three pieces of wooden broom handle, each about 3 ft. (1 m) long

3 Secure the second pole between two knots in the same way, leaving 16 to 20 in. (or 40 to 50 cm) of rope between the poles. Leave free an identical length of rope and tie the third pole in place.

4 Finally, tie the ends of the rope, so the three poles are fixed in an equal-sided rope triangle.

5 Now tie the other ends of the poles together in just the same way—but this time, make the sides of the triangle a little shorter: 12 to 16 in. (or 30 to 40 cm) long.

6 After all three poles are roped together at both ends, you are ready to set up your stool. It can be tricky, so you'll probably need to ask a friend to help you. Pull all three poles apart so that the rope triangles are stretched out. Stand the larger triangle on the floor, then twist the smaller triangle around until all three poles cross in one place. Now, rest a board on top and take a seat!

Tension and compression
When you sit on the stool, your weight pushes down on the poles and **compresses** them. The **tension** in the ropes keeps the poles from sliding and keeps the stool rigid.

Experimenting with shapes
Think about the strength of different shapes. Make a stool frame in the shape of a square using 4 small pieces of wood. Is it easy or difficult to push or twist it out of shape? Now try the same thing with a triangle made the same way. What happens?

Find a book about engineering or bridge building. What shape do you see used in construction? Why is this?

World of Plants

Without plants, Earth would be lifeless. The oxygen that animals breathe comes from plants. Most of our food comes from plants and from animals that eat plants or plant-eating animals.

We build shelter out of plants and make many useful products from them, such as medicine, paper, and cloth.

Botanists, the scientists who study plants, try to answer many questions. What do plants have in common? How are they different? Why are plants important to us after they die?

Start studying the plants in your area. Visit your local nursery or nature center to learn more about the amazing attributes and variety of plants. Then, in this chapter, explore *Classifying plants, Plant growth, Decay,* and *Natural dyes.* Each project is a starting point for answering basic questions about plants. Some of these questions follow. Can you think of others? Each one could lead to an exciting science fair project.

Plant growth

Does the quality of a plant vary when grown from a seed, bulb, or cutting?

How do the quality and quantity of water, light, and soil affect the germination of seeds? The taste of an edible plant?

What characteristics can you use to measure the growth of plants?

Do the color and healthful appearance of a plant play a part in its growth?

How do natural predators affect the growth of a plant? How does trimming a plant's leaves affect its growth? Trimming its roots?

Decay

What are the best methods for preventing decay of your fruits, dairy products, and other foods?

What conditions in the environment (or in the refrigerator) promote decay?

Can you determine through tests what particular molds decay organic matter in your home? Where do they originate?

How does a compost pile speed up the decay process? How do the contents change over time? What factors in the environment create the greatest changes?

Classifying plants

What is the best method for collecting and preserving plants to use them for further study?

How do members of the same species vary from one habitat to another?

How does the mix of plant species vary from one natural habitat to another?

Natural dyes

What natural dyes produce the most lasting color?

How do natural dyes from plants differ from human-made dyes? Can you test them and compare their chemicals?

Do some materials accept dye better than others? Why or why not?

Botanists collect plant **specimens** to help them study plants more closely. By collecting, **cataloging**, and comparing natural objects, botanists try to discover differences and similarities between species of plants. The collection and study of such **data** are vital to all science.

Discover!
Ask an adult to help you collect your own specimens. Try to collect things such as flowers, seeds, and leaves. Then make a box to display and store your collection.

You will need
cardboard	glue or tape, utility knife
ruler, pencil	toothpicks
sawdust, cotton, or	craft sticks
tissue paper	colored paper

1 Decide what size you want your box. Out of cardboard, cut a flat shape like the one on the right.

2 Draw lines on your box to mark the exact position where the dividers will go.

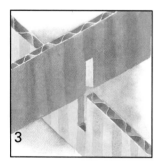

3 Make the dividers by cutting strips the same height as the box. Make slots by cutting a slit halfway up one divider and halfway down the other and then fit the dividers together.

4 Fold up the sides of the box. Tuck in the corner flaps and glue or tape them in place. Push the dividers into place in the box.

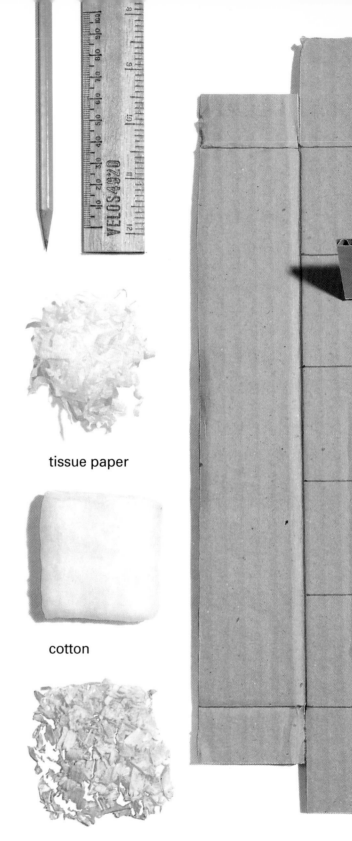

tissue paper

cotton

sawdust

5 Fill the sections of the box with sawdust, cotton, or tissue paper. Use glue to keep them in place. Then add your collection.

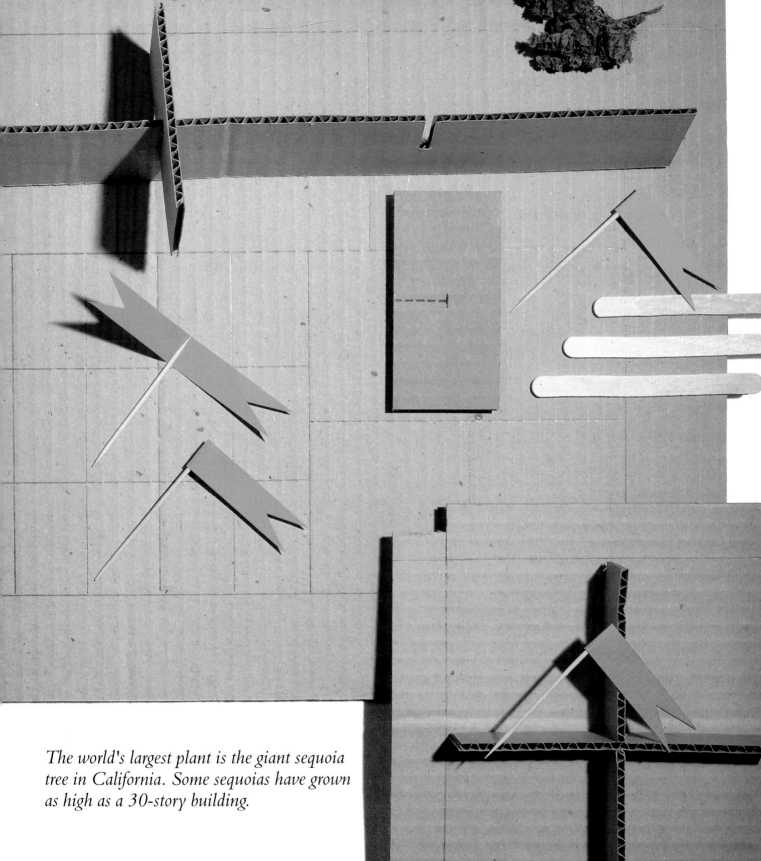

The world's largest plant is the giant sequoia tree in California. Some sequoias have grown as high as a 30-story building.

Labeling
Sort your specimens into different categories such as seeds, flowers, leaves, or bark. Try to find out the names of your specimens by looking for them in books. Label each specimen with both its scientific name and its common name, if it has one. To make the label, write the names on a craft stick or make a flag from colored paper and a toothpick. Tape or glue the label to the box.

Preserving

Preserve specimens by hanging them upside down in a warm, dry place, or by pressing them in a book. Describe how the plant looks when it is growing; its appearance will change as it dries. Draw or photograph it, if your science fair does not allow plants or plant specimens at the fair. Photograph the process and your collection.

Analyzing your collection

Collect specimens from different areas. Before you go, be sure you know which in the area are poisonous and should be avoided; bring along pictures and descriptions. Don't take specimens from protected lands.

While in the field, take notes of the areas from which you gather specimens. What is the terrain? Hilly? Flat? Grassy? Wooded? How would you describe the soil? Rocky? Sandy? Muddy? Are there lots of plants and animals? Are they the same kind? Take a photo or make a sketch of each area. Catalog your specimens and examine them. How do they vary from one area to another? Which are healthy in which areas? Research the plants in an area. What does their presence tell you about the local environment? About the health and composition of the soil?

54 Plant growth

Most plants make their own food, from minerals and salts in the soil, sunlight, and carbon dioxide. But not all plants grow at the same rate. Some have a life cycle of only one year, such as corn. Others, such as trees, take years to mature and continue growing for centuries.

Discover!

Experiment with different ways to grow plants from seeds, bulbs, and cuttings. Keep track of their growth, using the methods described on pages 56 and 57. Create a control and test one variable at a time. Try soaking the seeds or bulbs in water or nutrients before planting them. Does more water or more sunlight produce a healthier plant? Does it produce more or larger leaves? More, larger, or tastier fruit? What makes a plant grow faster? Does trimming the stems or roots impact growth?

watercress seeds

sweet corn

broad beans

You will need

a shallow dish	paper towels
watercress seeds	small stones
potting soil	a bulb
a cutting from a plant	
flower pots and saucers	
a glass of water	

1 To grow the watercress seeds, cut a circle of paper towel to fit the bottom of the shallow dish. Moisten it with water and sprinkle on the watercress seeds. Keep the paper towel moist and watch for the watercress to sprout.

▲ Record the size and appearance of your plants after two days.

▲ Record the size and appearance of your plants after four days.

daffodil bulb

hyacinth bulbs

2 To plant the bulb, put a layer of small stones in the saucer of your pot's base—this is to help the water drain out of the pot so that the bulb doesn't get waterlogged and soggy.

3 To root the cutting: Put the stem of the cutting into a glass of water. Check it every day. When it begins to develop roots, plant it in potting soil. Water it regularly.

Fill the plant pot with potting soil and bury the bulb in the center. The pointed end of the bulb should be about ½ inch (1.25 cm) below the surface of the soil.

Keep the soil moist and put the pot in a dark cabinet. Check it daily. When a green shoot breaks through, bring it out into the light. Bulbs should be watered little, but often.

To make a cutting, like the one pictured here, ▶ *cut a stalk off a healthy plant, just where it joins the main stem.*

Most green plants grow in two general directions. They grow *up* from the tips of their stems and *down* from the tips of their roots. Woody plants, such as trees and shrubs, grow up and down, too, but they also grow outward. Every year, they add a layer to their trunks, making them wider.

Plants grow because, like you, they have growth hormones. These hormones control the way a plant grows, making roots go down and stems go up. This up-and-down growth is called primary growth; width growth is known as secondary growth.

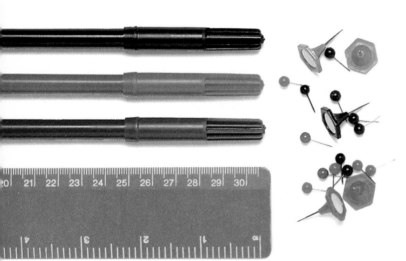

Discover!
Compare the growth of different types of plants, or test different variables on the same plant species. For example, what effect does humidity or heat have on a plant? Whatever variable you are testing, be sure all other conditions for your plants are the same. Measure your plants at regular intervals. Measure carefully and precisely. Take detailed notes of how each plant changes.

To measure growth, you will need
ruler	poster board
wooden sticks	thumbtacks
glue, tape	paint and paintbrush
graph paper	plain paper, newspaper

1 To measure root and stem growth: Make a measuring stick out of poster board as shown. Use a ruler to mark out inches or centimeters, starting with zero in the center and having the measurements run to both ends. The pointed side is used to measure the roots.

▼ *Plant a batch of watercress. Take one seedling out each day to measure it. Place the seed on the zero and keep a record of the root growth on one side, and the stem growth on the other.*

day 1

day 4

Experimenting with plant growth
Try growing a cutting from a plant you grow from a bulb. Record its growth. How does it compare to the growth rate of its parent plant? Is there any difference? Why or why not? You may want to plot the data on a chart or graph for easier comparison.

Compare plants grown in fertilizer to those grown without. How much fertilizer is required to affect growth rate? How much is too much? Be sure to perform a carefully controlled experiment along with ones that test your variable.

2 To measure stem growth: Put a tall wooden stick into each pot. Make labels out of poster board and glue them onto the stick to mark each plant's growth. Measure every week and record the growth. Try measuring different types of plants. Which plant grew more quickly than the others?

3 To measure leaf growth: At regular intervals, take a mature leaf from the same tree (springtime is best). Measure the leaf on graph paper as shown. Cover the underside of the leaf with thick paint. Place the leaf paint-side down on a piece of plain paper and cover it with newspaper. Gently rub the newspaper with a roller to press the ink onto the plain paper. After the paint dries, cut out the print and glue it onto the graph paper or poster board.

Each time you measure growth, also count and record the number of leaves and buds.

Even after plants die, they are still very important to the environment. Decaying plants return nutrients to the soil, making it rich and fertile for the next generation. Plants begin to **decompose** as soon as they are picked. The mold spores that cause decay creep into the plant (as its protective layer of skin starts to break down) and multiply very quickly. As food decays, it shrivels and becomes lighter, because the mold spores are eating it. Mold spores are **microorganisms**, which means they are so small that they cannot be seen without a microscope.

You will need

fruits, vegetables, kitchen knife   file box
camera or colored pencils   sturdy paper
rubber cement   plain paper

Be very careful with decayed food! Never taste it. And always wash your hands after touching it.

Discover!

The best way to study decay is to watch it happen!

1 Cut some fruits and vegetables. Leave them out and uncovered to spoil.

2 Take photos of them every few days or draw pictures of them. Record the conditions of the environment as well. Use rubber cement to attach your photos or drawings to sturdy paper along with your notes and labels. Store the cards in a file box.

orange

Experimenting with decay

Try leaving pieces of food in different places to test the length of time they take to decay. Leave some on a warm windowsill, some in a cool cabinet, and some outside.

Where do your samples decay most quickly? Least quickly? What conditions seem to preserve food? What conditions cause food to decay most rapidly? Why? Think about variables such as air temperature, humidity, and sunlight.

Are any of your samples moldy? What does this tell? Speculate on where the mold came from.

Research the history of food preservation. How is food preserved now? What are some of the advantages and disadvantages of modern methods of food preservation? Of older methods? Think about such issues as nutrient content, taste, and cost of food preserved in various ways versus unpreserved food.

▲ *A cover protects your files.*

▼ *Tabs help organize your file box.*

Bacteria are microorganisms. Some bacteria cause disease, fevers, and sore throats. Doctors have used the medicine penicillin to kill such bacteria.

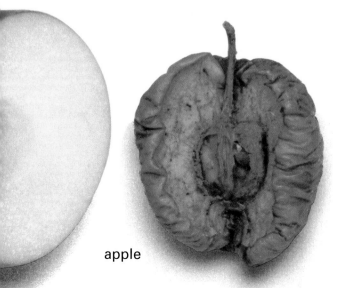

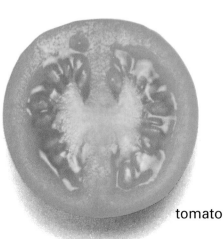

apple

tomato

You may have noticed that if you spill beet juice on a white T-shirt, it doesn't wash out easily. Beet juice is a natural dye. Today, most clothes are colored with human-made dyes, but for thousands of years before these were invented, people had been using natural dyes made from plants to bring color into their lives.

Discover!

Try extracting dye from onion skins and use it to tie-dye a square of cotton or wool.

To make dyes, you will need

large enamel or stainless steel pot
onion skins, muslin, clean marbles and pebbles
wooden spoon, white string, strainer
glass jars with lids, cotton, wool, water
alum (aluminum potassium sulphate)
cream of tartar, vinegar
spices, such as paprika, mustard powder,
 dried dill weed, chili powder, etc.

1 Peel the skin gently from a large brown onion. Put the skins into a piece of muslin, tie it up, and put it into a pan of cold water.

2 Wet the cotton or wool and put it in the pan. Bring the water to a gentle simmer and keep it simmering until the fabric is the color you want. It will take from 20 minutes to 3 hours. Stir it regularly.

3 Lift the fabric out of the pan with a wooden spoon and rinse it in clean, warm water until no more color runs out. Allow it to dry naturally.

Be careful!

Pans of boiling water are dangerous. Make sure your pan is big enough to hold the fabric and the water without spilling. Never try to boil water without an adult around to help you.

Making dyes colorfast

Some natural dyes are not colorfast, which means they will gradually wash out of the fabric if a substance called a **mordant** is not added. The mordant combines with the dye to make the color enter the fibers of the fabric and produce a color that will not fade.

Discover!

Try dyeing fabric using a mordant and spices.

1 Add 1 ounce (28.5 g) of one of the spices listed on page 60 to 1 quart (1 l) of water in an enamel pot. Bring it to a boil, then let it simmer for one hour. Add 2 teaspoons (10 ml) alum and ¼ tsp. (1.25 ml) cream of tartar. Stir to dissolve.

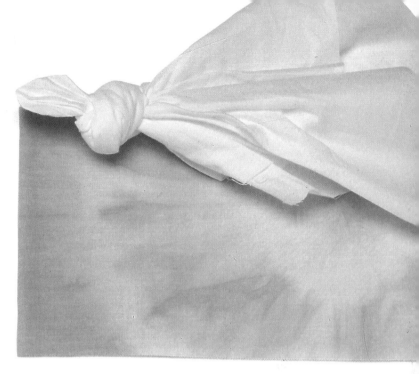

▲ *For your science fair display, make an*
▼ *exciting tie-dye pattern. Just tie a knot in the center of the fabric for a random effect (above). For a circular pattern (below), put a pebble or marble in the center of a cloth. Fold the cloth around it, and secure it with string. Use a series of marbles or pebbles.*

2 Add the cloth to the dye bath and simmer for another hour. Remove the pot from the heat and let the dye bath cool. Rinse the cloth until the water runs clear. Squeeze out excess water, and hang the cloth up to dry.

Experimenting with dyes and mordants

Try combining different dyes and mordants, or test natural dyes against synthetic ones. Set up a control, and make a chart and drawings of your results. Which are brighter? Fade faster? In the mordant recipe above, replace the alum and cream of tartar with ½ cup (or 120 ml) vinegar. What happens to the dye? Do colors change? Dye a cloth without a mordant, then wash the cloth. What happens? Leave cloth dyed with mordant and cloth dyed without mordant in the sun. What happens? Try using the same dye on different kinds of cloth.

Keeping the dye

You can use your dye again and again, as long as you keep it free from vegetable matter, which would quickly spoil it. Even if you have tied up your plants, roots, or skins securely in a muslin pouch, little particles will escape. When the dye has cooled, remove the muslin pouch and pour the dye through a strainer into clean glass jars with screw-top lids. Label the jars and keep them in a dark cabinet, as exposure to sunlight will cause the dyes to fade. As soon as the dyes start to look cloudy, throw them away.

▶ All the plants shown here will produce dyes that are naturally colorfast. Follow the directions on page 60, remembering to put the plant, root, skins, or powder into a muslin pouch. For the turmeric powder (a spice used in India for cooking), you may need a double layer of muslin. Note: The colors of the fabrics dyed with the yellow flowers of the goldenrod are in fact more green than yellow.

Among the most popular dyes are the intense yellow ones obtained from the stamens of the saffron crocus, the dark blue ones that come from the Indian indigo plant (which we know better as the color of blue jeans), and the orange-red of the henna plant, which is used today as a hair dye.

The ancient Phoenicians discovered that certain spiny snail shells, when crushed, produced a beautiful purple dye. It was very costly, and with the Romans in particular, the color became associated with wealth and power. To this day, kings and queens wear ceremonial robes of deep purple.

tea bags or loose tea

turmeric powder

avocado skins

beet juice

onion skins

goldenrod

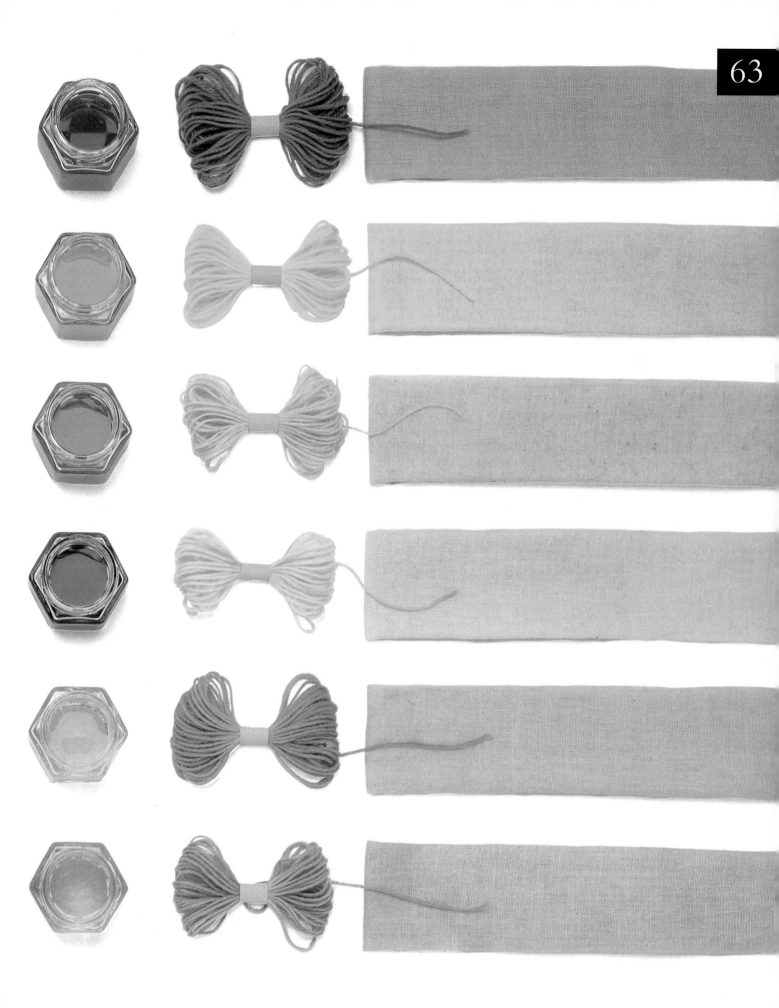

Sound

Vibrations are around us and within us—from air molecules bouncing off our eardrums, to our muscles shivering. We perceive vibrations in three ways: through sight, touch, and sound. Humans can hear sound produced by vibrations between about 20 and 20,000 hertz (vibrations per second). But cats can hear very high pitches, up to 65,000 hertz, and elephants often communicate below 20 hertz.

Scientists and engineers study many aspects of sound because sound is important to us in so many ways—communication, recreation (music), and science (for example, in predicting earthquakes). Try your skills as an acoustical engineer by exploring the following experiments: *Reflecting sound, Traveling sound, Recording sound,* and *Radio waves.* Each project is a starting point for answering basic questions about sound. Some of these questions follow. Can you think of others? Each one could lead to an exciting science fair project.

Reflecting sound

Do different mediums (materials) reflect sound in the same way or with the same volume?

What conditions affect whether you can hear some things drop, such as a piece of paper, a pin, and a feather?

Is it possible to make your room soundproof? For example, can you keep the "noise" from your stereo inside your room?

Radio waves

On a homemade radio, how can you receive the most radio stations?

What materials receive a transmission most clearly?

Does the size of a radio make a difference? Longer wires? A bigger tube?

What conditions are best for radio reception? How does weather affect it? Location?

Recording sound

Can you use an umbrella as a device to aid recording? Does using a different shape work better?

What else can be done to make recording quality better?

How does the shape of a microphone affect its recording ability?

Traveling sound

What shapes best harness sound waves to make them easily magnified and heard?

How do temperature, density, and elasticity of a medium affect the speed of sound?

Can you make a homemade telephone and wire it to use in your house as an intercom system?

Does distance make a difference in the quality of sound?

Sound waves can travel through all kinds of substances. Our ears often pick up sound waves traveling through air. Whales hear sound waves in water. And sound can even pass through a solid object, such as wood, if it is made to vibrate. However, sound doesn't pass very well through some materials. For example, if you shout at a brick wall, not much of the sound gets from the air through the wall and out the other side. Instead, the sound waves bounce back off the wall again.

Angles of bounce

Sound waves in air will bounce off a flat, solid object at the same angle as they hit it—just like a bouncing ball off a wall. If, however, the sound waves are bounced off a surface that is soft or bumpy, the waves will break up and fade away.

You will need

thick poster board   modeling clay
a craft knife   ruler
cardboard tubes
a tape recorder with microphone
a clock or watch with a very soft tick

Discover!

You can control the path of the sound waves by directing them along cardboard tubes. The tubes hold the sound together, making it louder because the sound waves can't spread out and get lost in the air around them.

1 Take four pieces of cardboard tube that are the same size. Cut three squares out of a piece of thick, smooth poster board.

2 Use modeling clay to secure the cardboard tubes and the squares of poster board in position as shown below. Each tube must be placed at exactly the same angle to the squares as the others have been.

3 Measure the distance in a straight line between one end of the zig-zag and the other.

4 Set up your clock or watch away from the tubes, and record it ticking across the distance you have measured. Does your microphone pick up any sound?

5 Now position the clock at one end of the zig-zag, and record the sound that comes out of the other end.

6 Try altering the position of the tubes, and record what happens.

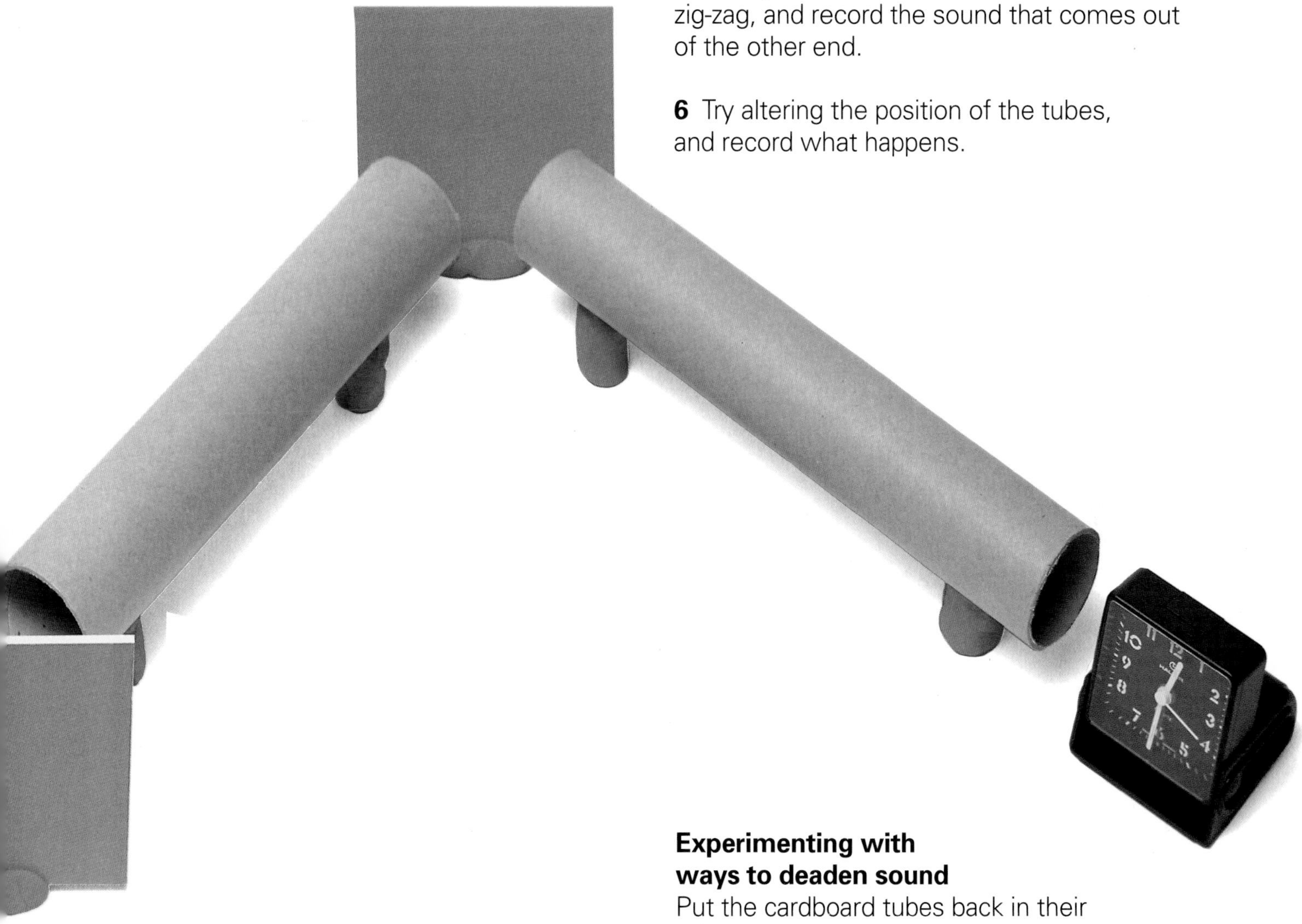

Architects use their knowledge of bouncing sound waves in many ways when they design new buildings. With the right floors, walls, and ceiling, a noisy restaurant can be made quieter. On the other hand, a concert hall is designed to reflect sound waves clearly to where the audience is sitting.

Experimenting with ways to deaden sound
Put the cardboard tubes back in their original zig-zag positions and then experiment with reflector cards made out of different materials. Cut some squares out of an old egg carton, so you can test the effect that a bumpy, uneven surface has on sound waves. Which surface reflects the most sound? Which reflects the least? Why? Research sound waves, and describe why the angle of the tubes is important. How do all of these things affect soundproofing? Why is soundproofing important?

Sound waves can be converted from sound energy into electrical energy and back again. That's how sounds can be transmitted through wires to the telephone. The sound waves are turned into electrical impulses, which travel through the telephone wire to the receiver–just as nerve pulses carry sounds from our ears to our brains.

Discover!

You can make a simple telephone using magnets and copper wire. When you speak into the paper cup, the paper drumhead **vibrates** and causes the magnet to bounce up and down inside a coil of copper wire. This movement produces pulses of electric current in the wire. At the other end of the wire, another copper coil, magnet, and drumhead convert the electrical pulses back into sound waves.

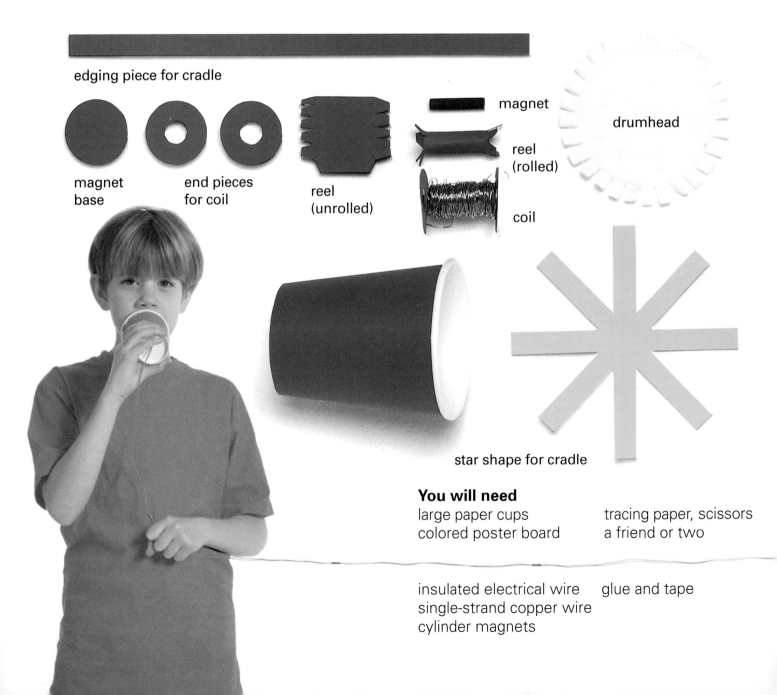

edging piece for cradle

magnet base

end pieces for coil

reel (unrolled)

magnet

reel (rolled)

coil

drumhead

star shape for cradle

You will need

large paper cups
colored poster board

tracing paper, scissors
a friend or two

insulated electrical wire
single-strand copper wire
cylinder magnets

glue and tape

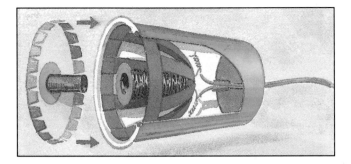

4 Now assemble a paper cup receiver. First, glue the coil into the cradle. Glue the cradle inside the paper cup as shown.

5 Cut a drumhead out of tracing paper. Glue the magnet to a small, circular magnet base cut from poster board and glue that onto the back of the paper drumhead. Then glue the drumhead in place. Check that the magnet fits neatly inside the reel, but has room to move freely up and down.

6 Repeat steps 1–5 to make the second receiver. For step 3, use the other end of the wire that you connected to the first copper coil.

7 When you talk, talk directly at the drumhead.

Experimenting with your telephone
Does distance and the length of the connecting wire affect sound quality? Does the size of the magnet or coil? Does the number of receivers?

1 First make a copper coil. Cut and glue the pieces of poster board as shown to make a reel that will fit neatly around a cylinder magnet. Wind a long piece of copper wire around the reel forty or fifty times.

▲ cylinder magnet attached to drumhead

▲ cradle

2 Cut a star shape and edging piece out of thick poster board as shown. Fold the spokes of the star shape and glue the edging piece to them to make a cradle.

3 Pass two equal pieces of insulated wire through the bottom of a paper cup and connect them to the end of the copper coil. Leave the connecting wire long enough to stretch at least across a large room.

Inside a tape recorder, sound waves are turned into electrical impulses. These are stored on the tape as a sequence of different magnetic blips.

Discover!

A microphone is a kind of electric ear. It turns sound waves into electric signals. However, it may have trouble picking up sounds not made close by. Try to improve a simple microphone by using an umbrella to reflect sound waves!

1 Begin by lining your friends up in a row. Stand facing them with the microphone, and ask them to sing a song together.

2 Now tape the microphone to the handle of an umbrella, as shown on the next page. Try to make the same recording again, standing in exactly the same place. Are the voices more evenly balanced when you play back the tape?

3 Take your equipment outside. Listen for birds singing, and then try to record them by pointing the microphone in their direction.

You will need

a tape recorder
a microphone
tape
three or four friends who will sing

blank cassette tapes
earphones
an umbrella

4 Now try using the umbrella to record the birds. You can make a collection of your recordings and keep a record of when and where you taped the different birds.

Experimenting with recording sound

Try this experiment again with umbrellas of different sizes. Do you think a larger or a smaller umbrella will work best? Why? What do your test results show? Try using other objects to collect sound waves—for example, a cardboard box. How do you predict that the size and shape of the object you choose will affect your recording? Why? Try several different containers and see if your predictions are correct. What other variables can you test? Consider the shape and material of the microphone.

Research acoustics (the science of sound). Read about how rooms are designed to amplify sound, or to deaden it. Based on your experiments, what shapes do you think would be used in a concert hall?

Radio waves are rays of electromagnetic energy, which we cannot see or feel. They travel at the speed of light through air, solid objects, and even empty space. At a radio transmitting station, sounds are first turned into electrical impulses, and then sent out as radio waves. A radio receiver picks up these waves and turns them back into sounds again.

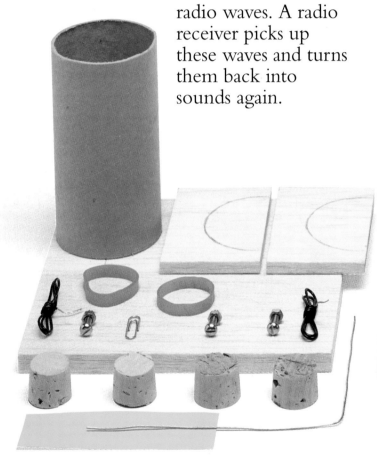

Discover!
Make your own simple radio receiver and see which radio signals you can tune into. Before you build it, speculate on what conditions will lead to the best reception—for example, weather and location. Explain why you think so. Build your receiver and test your theories.

1 Ask an adult to help you use the utility knife to cut a square piece of balsa wood for the baseboard.

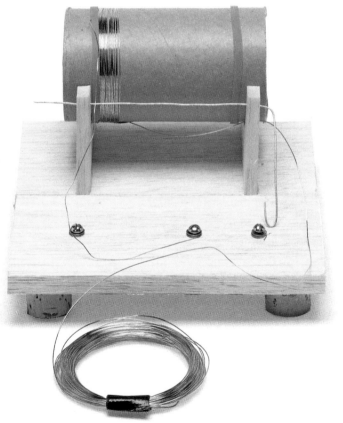

2 Cut half circles out of two other pieces of wood as shown above. Glue them into position on the base so that they will support the cardboard tube. Glue the corks at the corners of the bottom of the base as feet.

3 Drill four holes on top of the base and screw bolts into three of them as shown. Put a washer between each bolt head and the base. Secure the bolts beneath the board with nuts.

You will need

four corks	steel wire
an awl	a paper clip
wood glue	earphones
balsa wood	utility knife
a cardboard tube	thick rubber bands

bare copper wire
a diode, from a hobby shop
nuts, bolts, and washers
single-strand electrical wire (insulated)

4 Put the plug of the earphones into the remaining hole.

5 Wind the copper wire around the cardboard tube as tightly as possible. The turns should touch one another, without actually overlapping. Leave the ends of the wire loose, the left one long and the right one short. Secure the coil with thick rubber bands at both ends.

6 Secure the longer end of the copper wire beneath the left-hand bolt and around the earphone plug. Attach the very end of the wire to the middle bolt.

▼ *The diode directs the electromagnetic impulses picked up by the* **antenna,** *forcing them in one direction along the wire and into the earphones.*

7 Secure the diode between the middle bolt and the unconnected outer bolt.

8 Push the steel wire through the balsa wood supports in front of the copper coil, and bend it around to connect it to the diode as shown.

9 Attach a piece of insulated wire to each of the outer bolts. To make an antenna, take the wire connected to the diode, and fasten it to a long metal object, such as a radiator.

10 To operate the radio, slip a paper clip onto the steel wire, twist it forward, and move it along the copper coil. Grasp the unconnected end of insulated wire and put on the earphones.

Experimenting with radio waves
How many radio stations can you pick up with your receiver? In your project journal, note the reception for each. What variables can you change to receive more stations? Better reception? Consider the wires, the size of the tube, your location, and the weather.

74 Scope and Sequence

The projects in this book give students the opportunity to practice a wide variety of skills needed to excel in science, school, and lifelong learning. Educators agree that carrying out science projects helps develop important personal qualities, such as confidence, perseverance, and creativity—skills that are invaluable in almost every aspect of life.

When used in conjunction with the procedures described in *The Best of the Fair* and the questions posed in the section openers, all the projects in this book emphasize fundamental skills related to the scientific method—such as discussing

	Solar power	Living in space	Finding stars	Wind direction	Shifting plates	Rocks	Soil
Scientific literacy							
understanding and relating systems		✔		✔			
understanding scientific change	✔				✔	✔	✔
understanding structure and function		✔					
understanding processes		✔	✔		✔	✔	
Problem-solving skills							
thinking critically		✔	✔	✔		✔	
identifying causes and effects	✔	✔	✔	✔	✔		✔
defining a problem	✔	✔				✔	✔
Using scientific tools							
using tools properly	✔		✔	✔		✔	✔
modeling	✔	✔			✔		✔
measuring	✔	✔	✔	✔			✔
Mathematical skills							
recording	✔	✔		✔		✔	✔
constructing geometric figures			✔	✔			
measuring angles	✔						
understanding proportions		✔	✔		✔		✔
Logic skills							
classifying				✔	✔	✔	✔
comparing/contrasting	✔	✔	✔	✔	✔	✔	✔
controlling variables	✔	✔				✔	
Observational skills							
using senses		✔	✔	✔		✔	
gathering information	✔	✔	✔		✔	✔	✔
Communication skills							
describing	✔	✔	✔	✔	✔	✔	✔
working in a team							
labeling			✔		✔	✔	✔

scientific topics, questioning, predicting outcomes, hypothesizing, planning, testing a hypothesis, providing detailed facts and supporting evidence, and drawing and sharing conclusions. Use this *Scope and Sequence* chart to analyze additional skills that are highlighted by the activities and ideas in this book. Scan across the chart to find projects that emphasize a particular skill, or down the chart to see which skills are highlighted in a particular project.

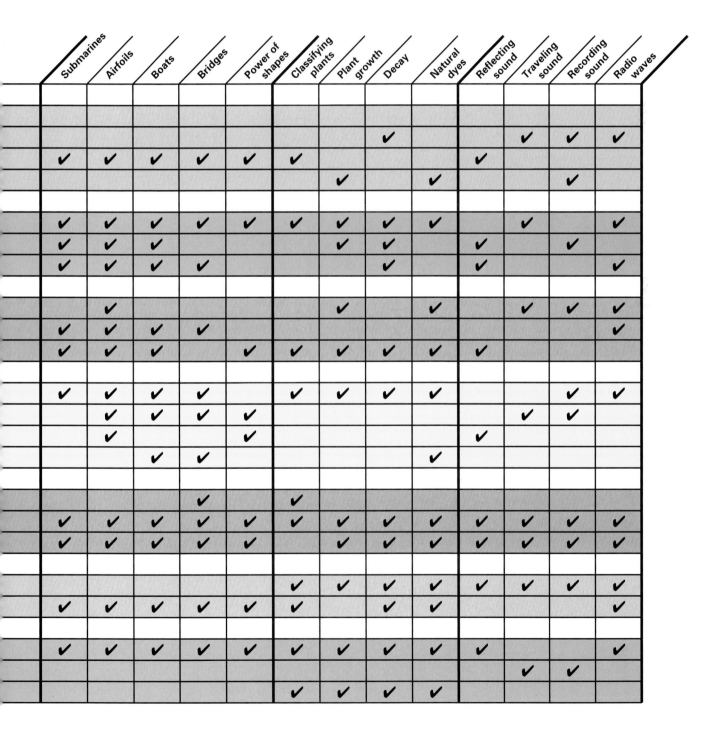

Submarines	Airfoils	Boats	Bridges	Power of shapes	Classifying plants	Plant growth	Decay	Natural dyes	Reflecting sound	Traveling sound	Recording sound	Radio waves
							✔			✔	✔	✔
✔	✔	✔	✔	✔	✔				✔			
						✔		✔			✔	
✔	✔	✔	✔	✔	✔	✔	✔	✔		✔		✔
✔	✔	✔				✔	✔		✔		✔	
✔	✔	✔	✔				✔		✔			✔
	✔					✔		✔		✔	✔	✔
✔	✔	✔	✔									✔
✔	✔	✔		✔	✔	✔	✔	✔	✔			
✔	✔	✔	✔		✔	✔	✔	✔			✔	✔
	✔	✔	✔	✔						✔	✔	
	✔			✔					✔			
		✔	✔					✔				
			✔		✔							
✔	✔	✔	✔	✔	✔	✔	✔	✔	✔	✔	✔	✔
✔	✔	✔	✔	✔		✔	✔	✔	✔	✔	✔	✔
					✔	✔	✔	✔	✔	✔	✔	✔
✔	✔	✔	✔	✔	✔			✔	✔			✔
✔	✔	✔	✔	✔	✔	✔	✔	✔	✔			✔
										✔	✔	
					✔	✔	✔	✔				

Acid *(AS ihd)* A type of chemical. Foods that contain acids taste sour or sharp.

Alkali *(AL kuh ly)* A type of chemical. Alkalis can combine with acids to make salts. Some plants grow well in alkaline soil.

Antenna *(an TEHN uh)* A long piece of wire or metal that picks up radio waves. Radios and televisions use antennas.

Ballast *(BAL uhst)* The extra weight added to a ship to give it stability or to make it float evenly.

Calcite *(KAL syt)* A mineral found in limestone. Calcite deposits form stalagmites and stalactites. Calcite is used in cement, paint, and glass.

Carbon dioxide *(KAHR buhn dy AHK syd)* A colorless gas with no smell. It is an important part of the air.

Cataloging *(KAT uh lawg ihng)* Organizing, listing, and storing information so that it can be found easily.

Compression *(KUHM prehsh uhn)* A force that squeezes things together. Compression squashes a cushion when you sit on it.

Data *(DAY tuh)* Information that has been gathered in an organized way.

Decompose *(DEE kuhm POHZ)* To decay or rot. Dead plants decompose with the help of bacteria.

Density *(DEHN suh tee)* The weight of a material in relation to its volume. For example, a lead ball weighs more than a plastic ball the same size. This is because lead has a higher density.

Drag *(drag)* The air resistance acting on airplanes, or the water resistance on boats.

Energy *(EHN uhr gee)* The ability to do work. People use energy stored in their muscles to move loads. Engines use the energy stored in fuels.

Engineers *(EHN juh NEERZ)* People who use scientific knowledge to invent, design, and make things such as roads, bridges, and buildings.

Environment *(ehn VY ruhn muhnt)* The conditions, such as soil, food, and climate, that surround a plant or animal. Every living thing needs the right environment to survive.

Evaporate *(ih VAP uh rayt)* To change from a liquid into a gas. When water evaporates, it becomes steam or water vapor.

Experiments *(ehk SPEYHR uh mehnts)* Tests done to confirm or disprove scientific ideas.

Galaxy *(GAL uhk see)* A group of billions of stars in space. There are billions of galaxies in the universe.

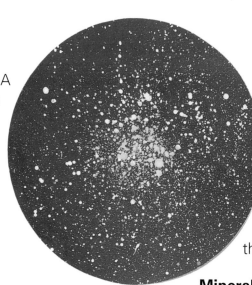

Geology *(jee AHL uh jee)* The study of Earth's structure, especially its rocks and minerals.

Hull *(huhl)* The main body of a vessel, the part that floats on the water.

Keel *(keel)* A wooden beam or steel girder running along the bottom of a ship's hull.

Latitude *(LAT uh tood)* The distance measured in degrees north or south from the equator. The equator is 0°. The North and South poles are at 90°.

Lattice *(LAT ihs)* A structure made by joining rods or beams into regular patterns. Lattices made from triangles are usually chosen for building because triangular shapes are strong.

Lift *(lihft)* A force that acts against gravity. Lift makes it possible for airplanes, airships, and balloons to rise in the air.

Microorganisms *(MY kroh OHR guh nihz uhmz)* Plants or animals that are too small to be seen by the human eye. Bacteria are microorganisms.

Milky Way *(MIHL kee way)* Our galaxy. The sun is just one of the billions of stars that make up the Milky Way.

Minerals *(MIHN uhr uhlz)* Substances, such as rocks or metals, that are not living things or made from living things. Sand and salt are minerals.

Mordant *(MOHR duhnt)* A chemical that combines with dye to produce a color that will not fade.

Photosynthesis *(FOH toh SIHN thuh sihs)* The process by which green plants use the energy from sunlight to change water and carbon dioxide into food.

Physics *(FIHZ ihks)* The branch of science that explores different kinds of energy and matter. Sound is one of the forms of energy studied in physics.

Prevailing winds *(prih VAY lihng wihndz)* Winds that usually blow from one direction.

Rudder *(RUHD uhr)* A large board fastened by hinges to the back or stern of a boat below the water. It is moved from side to side to steer the vessel.

Saturated solution *(SATCH uh RAY tihd suh LOO shuhn)* A liquid that has as much of another substance dissolved in it as it will hold. When water has dissolved as much salt as it can, the mixture is called a saturated solution.

Scientist *(SY uhn tihst)* A person who is an expert in a particular kind of study based on facts, such as biology, chemistry, physics, or astronomy.

Solar *(SOH luhr)* Anything referring to the sun, for example, a solar eclipse (eclipse of the sun) or solar energy (energy from the sun).

Sound waves *(sownd wayvz)* Vibrations of sound that travel through air and other materials. If we could see sound waves in air, they would look something like the ripples on a pond when a stone is thrown into the water.

Specimen *(SPEHS uh muhn)* A sample of a plant, animal, or rock that scientists use for experiments.

Stalactite *(stuh LAK tyt)* A long, thin deposit of limestone that hangs from the roof of a cave.

Stalagmite *(stuh LAG myt)* A long, thin deposit of limestone that grows up from the floor of a cave.

Strata *(STRAT uh)* Layers, usually of rock or soil.

Subsoil *(SUHB soyl)* The bottom layers of soil in the ground, nearest to the solid rock.

Tectonic plates *(tehk TAHN ihk playts)* The large sections of Earth's crust that hold the continents.

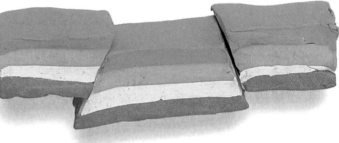

Tension *(TEHN shuhn)* A stretching force.

Theory *(THIHR ee)* An explanation based on learning and reasoning. Scientific theories usually are tested by doing experiments before they are said to be true.

Ties *(tyz)* Strings or rods that prevent two parts of a structure from separating.

Tiller *(TIHL uhr)* A long handle used for steering a boat. It is connected to the rudder.

Topsoil *(TAHP soyl)* The layer of soil nearest the surface of the ground. It contains the finest (smallest) soil particles.

Vibrate *(VY brayt)* Move back and forth very quickly. A guitar string vibrates if you pluck it.

Index

Find Out More

There are so many helpful resources available. Be sure to use some of them as you develop your science fair project.

For ideas and information about your topic, check out: books; computer programs; magazines, including children's, scientific, and specialty. Also check out libraries, associations, museums, and the Internet.

For project supplies, check out: your garage and basement; local businesses (they may donate new or leftover materials); craft supply stores; hobby shops; hardware stores; and other specialty stores.

Just a few of the many CD-ROM's and books about science:

Air: Simple Experiments for Young Scientists by Larry White (Millbrook Press, 1995)

The Art of Science by Joseph J. Carr (HighText, 1992)

Forces and Motion on CD-ROM for DOS/MAC/Windows (Science for Kids, 1994)

How to Excel in Science Competitions by Melanie J. Krieger (Watts, 1991)

Interfact: Electricity and Magnetism book and CD-ROM/floppy for MAC/Windows by Margaret Whalley (World Book, 1997)

Interfact: Solar System book and CD-ROM/floppy for MAC/Windows by Ian Graham (World Book, 1997)

MAKE it WORK! Electricity by Alexandra Parsons (World Book, 1997)

MAKE it WORK! Machines by David Glover (World Book, 1997)

MAKE it WORK! Time by David Glover (World Book, 1996)

Making and Using Scientific Models by Robert Gardner and Eric Kemer (Watts, 1993)

Science Fair Projects: The Environment by Bob Bonnet and Dan Keen (Sterling, 1995)

Science Fair Workshop by Marcia Daab (Fearon, 1990)

Science Through Art: Color by Hilary Devonshire (Watts, 1991)

The Way It Works: Air by Philip Sawain (New Discovery, 1992)

The Way Things Work on CD-ROM for Mac and Windows (Dorling Kindersley, 1994)

The World Book Encyclopedia on CD-ROM for DOS/MAC/Windows (updated yearly) contains science articles and projects.

Young Scientist series by World Book